T0211922

Fundamentals of Respiratory Sounds and Analysis

Fundamentals of Respiratory Sounds and Analysis

Zahra Moussavi

ISBN: 978-3-031-00489-6 paperback

ISBN: 978-3-031-01617-2 ebook

DOI: 10.1007/978-3-031-01617-2

A Publication in the Springer series
SYNTHESIS LECTURES ON BIOMEDICAL ENGINEERING #8
Series Editors: John D. Enderle, University of Connecticut

ISSN 1930-0328   Print
ISSN 1930-0336   Electronic

First Edition

10 9 8 7 6 5 4 3 2 1

Fundamentals of Respiratory Sounds and Analysis

Zahra Moussavi
University of Manitoba
Winnipeg, Manitoba,
Canada

SYNTHESIS LECTURES ON BIOMEDICAL ENGINEERING #8

ABSTRACT

Breath sounds have long been important indicators of respiratory health and disease. Acoustical monitoring of respiratory sounds has been used by researchers for various diagnostic purposes. A few decades ago, physicians relied on their hearing to detect any symptomatic signs in respiratory sounds of their patients. However, with the aid of computer technology and digital signal processing techniques in recent years, breath sound analysis has drawn much attention because of its diagnostic capabilities. Computerized respiratory sound analysis can now quantify changes in lung sounds; make permanent records of the measurements made and produce graphical representations that help with the diagnosis and treatment of patients suffering from lung diseases. Digital signal processing techniques have been widely used to derive characteristics features of the lung sounds for both diagnostic and assessment of treatment purposes.

Although the analytical techniques of signal processing are largely independent of the application, interpretation of their results on biological data, i.e. respiratory sounds, requires substantial understanding of the involved physiological system. This lecture series begins with an overview of the anatomy and physiology related to human respiratory system, and proceeds to advanced research in respiratory sound analysis and modeling, and their application as diagnostic aids. Although some of the used signal processing techniques have been explained briefly, the intention of this book is not to describe the analytical methods of signal processing but the application of them and how the results can be interpreted. The book is written for engineers with university level knowledge of mathematics and digital signal processing.

KEYWORDS

respiratory system, ventilation, respiratory sound analysis, lung sound, tracheal sound, adventitious sounds, respiratory sound transmission, symptomatic respiratory sounds

Contents

CHAPTER 1

Anatomy and Physiology of Respiratory System

1.1 OVERVIEW

The primary function of the respiratory system is supplying oxygen to the blood and expelling waste gases, of which carbon dioxide is the main constituent, from the body. This is achieved through breathing: we inhale oxygen and exhale carbon dioxide. Respiration is achieved via inhalation through the mouth or nose as a result of the relaxation and contraction of the diaphragm. The air, in essence oxygen, then passes through the larynx and trachea to enter the chest cavity. The larynx, or voice box, is located at the head of the trachea, or windpipe. In the chest cavity, the trachea branches off into two smaller tubes called the bronchi, which enter the hilus of the left and right lungs. The bronchi are then further subdivided into bronchioles. These, in turn, branch off to the alveolar ducts, which lead to grape-like clusters called alveoli found in the alveolar sacs. The anatomy of the respiratory system is shown in Fig. 1.1. The walls of alveoli are extremely thin (less than 2 μm) but there are about 300 millions of alveoli (each with a diameter about 0.25 mm). If one flattens the alveoli (in an adult), the resulted surface can cover about 140 m^2.

The lungs are the two sponge-like organs which expand with diaphragmatic contraction to admit air and house the alveoli where oxygen and carbon dioxide diffusion regenerates blood cells. The lungs are divided into right and left halves, which have three and two lobes, respectively. Each half is anchored by the mediastinum and rests on the diaphragm below. The medial surface of each half features an aperture, called a hilus, through which the bronchus, nerves, and blood vessels pass.

When inhaling, air enters through the nasal cavity to the pharynx and then through the larynx enters the trachea, and through trachea enters the bronchial tree and its branches to reach alveoli. It is in alveoli that the exchange between the oxygen in the air and blood takes place through the alveolar capillaries. Deoxygenated blood is pumped to the lungs from the heart through the pulmonary artery. This artery branches into both lungs, subdividing into arterioles and metarterioles deep within the lung tissue. These metarterioles lead to networks of smaller vessels, called capillaries, which pass through the alveolar surface. The blood diffuses

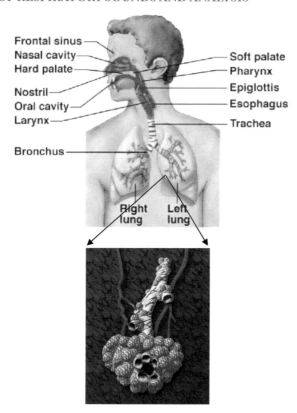

Frontal sinus
Nasal cavity
Hard palate
Nostril
Oral cavity
Larynx
Bronchus
Soft palate
Pharynx
Epiglottis
Esophagus
Trachea
Right lung
Left lung

FIGURE 1.1: Anatomy of the respiratory system (top view); the zoomed in picture of a bronchiolus branch and alveolar ducts (bottom view)

waste carbon dioxide through the membranous walls of the alveoli and takes up oxygen from the air within. The reoxygenized blood is then sent through metavenules and venules, which are tributaries to the pulmonary vein. This vein takes the reoxygenized blood back to the heart to be pumped throughout the body for the nourishment of its cells.

Ventilation is an active process in the sense that it consumes energy because it requires contraction of muscles. The main muscles involved in respiration are the diaphragm and the external intercostal muscles. The diaphragm is a dome-shaped muscle with a convex upper surface. When it contracts it flattens and enlarges the thoracic cavity. During inspiration the external intercostal muscles elevate the ribs and sternum and hence increase the space of the thoracic cavity by expanding in the horizontal axis. Simultaneously, the diaphragm moves downward and expands the thoracic cavity space in the vertical axis. The increased space of the thoracic cavity lowers the pressure inside the lungs (and alveoli) with respect to atmospheric pressure. Therefore, the air moves into lungs. During expiration, the external intercostal muscles and diaphragm relax the thoracic cavity which is restored to its preinspiratory volume. Hence, the pressure in the lungs (and alveoli) is increased (becomes slightly positive with respect to atmospheric pressure) and the air is exhaled. At low flow rate respiration, i.e., 0.5 L s^{-1} when lying on ones back, almost all movement is diaphragmatic and the chest wall is still. At higher flow rates, the muscles of the chest wall are also involved and the ribs move too. Different people breathe differently in terms of using the diaphragm to expand the lungs or the chest wall muscles. For instance, breathing in children and pregnant women is largely diaphragmatic. Without going through the pulmonary physiology in detail, it is necessary to introduce a few pulmonary parameters that will be referred to when we discuss the lung sound analysis.

1.2   VENTILATION PARAMETERS

Lung Volumes

a) *Tidal Volume (TV)*. It is the volume of gas exchanged during each breath and can change as the ventilation pattern changes, and is about 0.5 L.

b) *Inspiratory reserve volume (IRV)*. It is the maximum volume that can be inspired over and beyond the normal tidal volume and is about 3 L in a young male adult.

c) *Expiratory reserve volume (ERV)*. It is the maximum volume that can still be expired by forceful expiration after the end of a normal tidal expiration and is about 1.1 L in a young male adult.

d) *Residual Volume (RV)*. It is the volume remaining in the lungs and airways following a maximum expiratory effort and is about 1.2 L in a young male adult. Note that lungs cannot empty out completely because of stiffness when compressed, and also airway collapse and gas trapping at low lung volumes.

Capacities: Combined Volumes

a) *Vital capacity (VC)*. It is the maximum volume of gas that can be exchanged in a single breath: $VC = TV + IRV + ERV$.

b) *Total lung capacity (TLC)*. It is the maximum volume of gas that the lungs (and airways) can contain: $TLC = VC + RV$.

c) *Functional residual capacity (FRC)*. It is the volume of gas remaining in the lungs (and airways) at the end of the expiratory phase: $FRC = RV + ERV$. We normally breathe above the FRC volume.

d) *Inspiratory capacity (IC)*. It is the maximum volume of gas that can be inspired from the end of the expiratory phase: $IC = TV + IRV$.

Minute ventilation is the total flow of air volume in/out at the airway opening (mouth). Hence, Minute Ventilation $=$ Tidal Volume $\times$ Respiratory Rate.

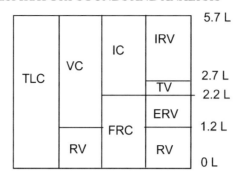

FIGURE 1.2: Volumes diagram

Dead space is the volume of conducting airways where no gas diffusion occurs. Fresh air entering the dead space does not reach alveoli, and hence does not mix with alveolar air. It is about 150 mL, which is about 30% of the resting tidal volume.

Fig. 1.2 shows a rough breakdown of these lung volumes. The vital capacity (VC) and its components can be measured using pulmonary function testing known as spirometry (Fig. 1.3), which involves inhalation of as much air as possible, i.e., to TLC, and maximally forcing the air out into a mouthpiece and pneumotachograph. Spirometry is the standard method for measuring most relative lung volumes. However, it cannot measure absolute volumes of air in the lung, such as RV, TLC, and FRC.

The most common approach to measure these absolute lung volumes is by the use of whole-body plethysmography (Fig. 1.4). In body plethysmography, the patient sits in an airtight

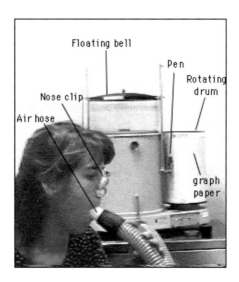

FIGURE 1.3: Spirometry

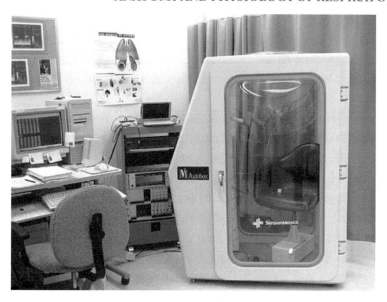

FIGURE 1.4: Plethysmography, Respiratory Lab, University of Manitoba

chamber and is instructed to inhale and exhale to a particular volume (usually FRC) and then a shutter drops across his/her breathing tube. The subject breathes in and out across the closed shutter (this maneuver feels like panting), which causes the subject's chest volume to increase and decompresses the air in the lungs. This increase in chest volume reduces the chamber volume; hence, increases the pressure in the chamber. Since we know the initial pressure (P_1) and volume of the chamber (V_1) and also the pressure of the chamber after the breathing maneuver of the subject (P_2), using Boyles law, $P_1 V_1 = P_2 V_2$, we can compute the new volume of the chamber at the end of the respiratory effort of the patient (V_2). The difference between these two volumes is the change of the chamber volume during the respiratory effort, which is equal to the change in volume of the patient's chest:

$$V_2 - V_1 = \Delta V_p = \text{Change in patient's chest volume.}$$

Now, we use Boyle's law again to find the initial volume of the patient's lung at the time when the shutter was closed. Let V_i be the initial lung volume (unknown), P_m be the pressure at the mouth (known), V_{ins} be the inspiratory volume of the chest (the unknown value) plus the change in the volume that we computed above, and $P_{\text{m-ins}}$ be the pressure at the mouth during the inspiratory effort (known). Using Boyle's law again, we can compute the initial volume of the lung when the shutter was closed:

$$V_i P_m = \left(V_i + \Delta V_p \right) P_{\text{m-ins}} \Rightarrow V_i = \frac{\Delta V_p P_{\text{m-ins}}}{P_m - P_{\text{m-ins}}}.$$

1.3   LUNG MECHANICS

The simplest and most common variables used to assess normal and altered mechanics of the respiratory system are *airway resistance* and *lung compliance*. Both of these parameters change in various disease states; hence, they are important parameters to assess the lung and respiratory system.

Airway resistance is analogous to blood flow in the cardiovascular system and also analogous to resistance in an electrical circuit while pressure and airflow are analogous to voltage and current in that circuit, respectively. Hence, one can conclude that the airway resistance can be measured as the change of pressure (voltage) to the flow (current). This measurement and relationship is true regardless of the type of flow. Recall that there are two types of airflows: laminar and turbulent. When the flow is low in velocity and passes through narrow tubes, it tends to be orderly and move in one direction; this is called laminar flow. For laminar flow, resistance is quite low and can be calculated by Poiseuille's law, which is then directly proportional to the length of the tube and inversely proportional to the fourth power of radius of the tube. Hence, the radius has a huge effect on the resistance when the flow is laminar; if the diameter is doubled the resistance will drop by a factor of 16.

On the other hand, when the flow is in high velocity, especially through an airway with irregular walls, the movement of flow is disorganized, perhaps even chaotic and makes eddies. In this case the pressure–flow relationship is not linear. Hence, there is no straightforward equation to compute airway resistance without knowing the pressure and flow velocity, and it can only be measured as the ratio of the change of pressure over the flow velocity. Airway resistance during turbulent flow is relatively much larger compared to laminar flow; a much greater pressure difference is required to produce the same flow rate as that of laminar flow.

Regardless of the type of flow, the airway resistance increases when the radius of the airway decreases. Therefore, at first glance at the respiratory system, it is expected that the larger airways, i.e., trachea, should have less resistance compared to that of smaller airways such as alveoli. However, it is opposite and can be explained by the electrical circuit theory. Recall that the bronchi tree has many branches in parallel with each other (i.e., parallel resistors); hence, the net effective resistance of the alveoli is much less than that of the larger airways, i.e., trachea. In fact, approximately 90% of the total airway resistance belongs to the airways larger than 2 mm.

Airway resistance is a very useful parameter as it can quantify the degree of obstruction to airflow in the airways. However, since the smallest airways get affected first by the development of an obstructive lung disease and also that most of the airway resistance appears in larger airways, the obstructive lung disease may exist without the symptoms of obstructive airways at least at early stages of the disease.

Compliance is a measure of lung stiffness or elasticity. Because of this inflatable property, the lung has often been compared to a balloon. For example, in fibrosis the lungs become

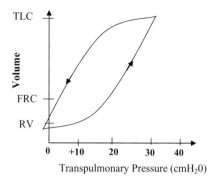

FIGURE 1.5: Pressure–volume hysteresis loop

stiff, making a large pressure necessary to maintain a moderate volume. Such lungs would be considered poorly compliant. On the other hand, in emphysema, where many alveolar walls are lost, the lungs would be considered highly compliant, i.e., only a small pressure difference inflates the lung.

Compliance is measured as the ratio of the change of volume over the change of pressure. However, the volume–pressure relationship is not the same during inflation (inspiration) and deflation (expiration); it forms a hysteresis loop (Fig. 1.5). The dependence of a property on past history is called hysteresis. Because of the weight and shape of the lung, the intrapleural pressure is less negative at the base than at the apex. Therefore, the basal lung is relatively compressed in its resting state but expands better than the apex on inspiration. It can be observed in Fig. 1.5 that the volume at a given pressure during deflation is always larger than that during inflation. Another important observation from the lung volume–pressure hysteresis curve is that the compliance changes with volume and actually it has a shape like an inverted bell with the peak near the FRC volume (Fig. 1.6). This implies that the lung has its highest compliance when we breathe at tidal flow (which is above the FRC volume); hence the minimum effort (pressure)

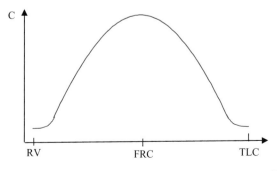

FIGURE 1.6: Lung compliance versus lung volume

is required for tidal breathing. One can correctly expect and experience that at higher volumes than FRC (higher flow rates) the lung becomes stiffer (less compliant) and breathing requires more effort (pressure).

In diseases such as fibrosis, the compliance is reduced and the lung becomes stiff. On the other hand, in a chronic obstructive pulmonary disease, i.e., emphysema, the alveolar walls degenerate; hence increasing the lung compliance.

In emphysema, the airways might be normal but because the surrounding lung tissue is progressively destroyed, it results in the obstruction to airflow and development of enlarged air sacs. Therefore, during inspiration they do not enlarge and on expiration they tend to collapse. Emphysema is a smoking-related disease that causes progressive obstruction of the airways and destruction of lung tissue. Because the airway is obstructed, more energy is required to ventilate the lungs; hence, the patient will experience shortness of breath.

Lung fibrosis, on the other hand, has the opposite effect of lung compliance change due to disease. In pulmonary fibrosis, the air sacs of the lung are replaced by fibrotic tissue; as the disease progresses, the tissue becomes thicker causing an irreversible loss of the tissue's ability to transfer oxygen into the bloodstream. By stiffening the lung tissue, airways in a fibrotic lung may be larger and more stable than normal. However, this does not mean that ventilation is easier in fibrosis. Even though the airway resistance may be smaller, the increased lung stiffness inhibits normal lung expansion making breathing very hard. For this reason, shortness of breath particularly with exertion is a common symptom in the patients with pulmonary fibrosis.

The lung tissues and airways become hyperresponsive in asthma, which results in reversible increase in bronchial smooth muscle tone and variable amounts of inflammation of bronchial mucosa. Because of the increased smooth muscle tone during an asthma attack, the airways also tend to close at abnormally high lung volumes, trapping air behind occluded or narrowed small airways. Therefore, asthmatic people tend to breathe at high lung volume in order to counteract the increase in smooth muscle tension, which is the primary defect in an asthmatic attack. Because these patients breathe at such high lung volumes and at that high volume based on the pressure–volume curve (Fig. 1.5) lung compliance is at its minimum (Fig. 1.6), they must exert significant effort to create an extremely negative pleural pressure, and consequently fatigue easily.

C H A P T E R 2

The Model of Respiratory System

Many researchers have worked on modeling the respiratory system both from merely scientific point of interest to understand how a biological system works and also for its plausible application for diagnostic purposes. The respiratory system also has a nonrespiratory function, which is vocalization. The sound generation of vocalization and that of respiration have similarities and also substantial differences. However, the vocal system has well-established models and theories while the respiratory sound generation and transmission is one of the controversial issues in respiratory acoustics due to its complexity. Since most of the respiratory sound transmission models are extensions of the acoustic model of the vocal tract (the part of the respiratory system between the glottis and the mouth/nasal cavity), in this book we start with describing a simple electrical T-circuit model to describe vocal tract acoustic properties for sound generation and transmission.

2.1   VOCAL TRACT MODEL

Sound is generated as a result of pressure change; hence it can be said that sound is a pressure wave propagated away from the source in a fashion similar to the wave as a result of dropping a stone into water. The pressure alternatively rises and drops as the air is compressed and expanded. That is why an object vibrates when a sound is loud enough.

Larynx is the source of pressure wave production which results in vocalization sound in human. Nasal cavity, lips, and tongue can also create sound as some animals, e.g., toothed whales, vocalize with structures in the nasal cavity. In human, the sound in larynx is generated by air moving past the vocal cords. The part of the vocal system inferior to the vocal cord is called subglottal and the part superior to that is called supraglottal. The constricted V-shaped space between the vocal cords is called the glottis. The larynx is constructed mainly of cartilages including the thyroid that is known as Adam's apple. The vocal cords are folds of ligaments between the thyroid cartilage in the front of neck and the arytenoid cartilages at the back. The arytenoid cartilages are movable and control the size of the glottis and hence produce different frequencies. The vocal cords are normally open to allow breathing and the passage of air into lungs. They close during swallowing as one of the many protection mechanisms during eating.

The vocal sounds are produced by opening and closure of the vocal cords or in other words by restricting the glottis.

Recalling that the sound generated by the vocal cords is in fact a pressure wave, it follows that the vocal sound has multiple frequencies: a fundamental frequency and a series of harmonic frequencies which are integer multiples of the fundamental frequency. The actual sound that is heard from the mouth is determined by the relative amplitude of each of the harmonic frequencies. The vocal tract acts as a bandpass filter that amplifies some frequencies and attenuates some others. Hence, it can be considered as a resonance chamber the shape of which determines the perceived pitch of the sound. It is the mass, tension, and length of the vocal cords that determine the frequency of the vibration. The vocal cords are typically longer and heavier in the male adults than in females; hence, male voices have a lower pitch than female voices. Note that the perceived pitch is not the real frequency of the sound. Pitch depends mainly on the frequency but in essence it is a subjective perception of the frequency by our ears and brain; hence, the same sound can be heard quite differently by two persons.

Despite the complexity of the human vocal tract with its many bends and curves, its main characteristics can reasonably be described by simple tube-like models and their analogous electrical models. The simplest model of the vocal tract is a pipe closed at one end by the glottis and open at the other end, the lips. Such a pipe has resonances at $f = \frac{nv}{4L}$, $n = 1, 3, 5, \ldots$, where v is the velocity of air and L is the length of the pipe. Two or more segment pipe models are proposed to model the vocal tract behavior for every vowel and other sounds production.

The length of the vocal tract is about 17 cm in adult men. Since this is fully comparable to the wavelength of sound in air at audible frequencies, it is not possible to obtain a precise analysis of the airway sound transmission without breaking it into small and short segments and considering the wave motion for frequencies above several hundred hertz. Practically, as mentioned before, the vocal tract is modeled as a series of uniform, lossy cylindrical pipes [1]. For simplicity, assume a plane wave transmission so that the sound pressure and volume velocity are spatially dependent only upon x. Due to the air mass in the pipe, it has an interance, which opposes acceleration. Because the tube could be inflated or deflated, the volume of air exhibits compliance. Assuming that the tube is lossy, there is viscous friction and heat conduction causing energy loss. With these assumptions, the characteristics of sound propagation in such a tube are described by a T-line electrical lossy transmission line circuit.

Having recalled the relations for the uniform, lossy electrical line, we want to interpret plane wave propagation in a uniform and lossy pipe in analogous terms. Note that the vocal tract is not really a homogenous, and hence a uniform, pipe. However, with this simplification assumption we can derive a simple model for a complex organ that represents the function of that organ reasonably well. Sound pressure, P, can be considered analogous to voltage and acoustic volume velocity, U, analogous to current. Then, the lossy, one-dimensional, T-line circuit represents the sinusoidal sound propagation with attenuation as it travels along the tube. In a smooth hard-walled tube the viscous and heat conduction losses can be analogously represented by I^2R and V^2G losses, respectively. As the equations below imply, the interance of the air mass is analogous to the electrical inductance, and the compliance of the air volume is analogous to the electrical capacitance. The parameters of this electrical model can be derived as follows [1].

The Acoustic L

The mass of air contained in the pipe with the length l is ρAl, where ρ is the air density and A is the area of the pipe. Recalling the second Newton's law and the relationship between force and pressure, the following equation can be derived to represent pressure in terms of a differential equation of volume velocity:

$$F = ma \Rightarrow PA = \rho Al\frac{\mathrm{d}u}{\mathrm{d}t} = \rho l\frac{\mathrm{d}U}{\mathrm{d}t} \Rightarrow P = \rho\frac{l}{A}\frac{\mathrm{d}U}{\mathrm{d}t}$$

$$\text{comparing with } V = L\frac{\mathrm{d}I}{\mathrm{d}t} \Rightarrow L_\mathrm{a} = \frac{\rho l}{A}.$$

Note that u is the particle velocity and $U = Au$ is the volume velocity. As shown in the above equations, the interance of air mass is analogous to electrical inductance.

Acoustic C

The air volume Adx experiences compression and expansions that follow the adiabatic gas law: $PV^\eta = \text{contant}$, where V and P are the total pressure and volume of the gas and η is the adiabatic constant. Differentiating the above equation gives

$$p\eta V^{\eta-1}\frac{\mathrm{d}V}{\mathrm{d}t} + V^\eta\frac{\mathrm{d}P}{\mathrm{d}t} = 0$$

$$\frac{1}{P}\frac{\mathrm{d}P}{\mathrm{d}t} = -\frac{\eta}{V}\frac{\mathrm{d}V}{\mathrm{d}t} = \frac{\eta}{V}U$$

$$\Rightarrow U = \frac{V}{P\eta}\frac{\mathrm{d}P}{\mathrm{d}t} \equiv C\frac{\mathrm{d}P}{\mathrm{d}t}$$

$$\therefore\ C = \frac{V}{P\eta}.$$

Compare the above equation for the volume velocity with $I = C\frac{\mathrm{d}V}{\mathrm{d}t}$. Recalling that the current, I, is analogous to the volume velocity, U, and the voltage, V, is analogous to pressure, P, we can derive the analogous acoustic C for compliance as $C = \frac{V}{P\eta}$. This equation for compliance is also in agreement with the measurement of compliance in pulmonary mechanics as mentioned in Section 1.3, which is measured as $\frac{\Delta V}{\Delta P}$.

Acoustic R

Acoustic R is defined as $R_a = \frac{lS}{A^2}\sqrt{\frac{\omega\rho\mu}{2}}$, where A and S are the tube area and circumference, respectively. ρ is the air density and μ is the viscosity coefficient.

Acoustic G

Acoustic G is defined as $G_a = Sl\frac{\eta-1}{\rho c^2}\sqrt{\frac{\lambda\omega}{2c\rho}}$, where c is the sound velocity, λ is the coefficient of heat conduction, η is the adiabatic constant, and c_p is the specific heat of air at constant pressure.

Having defined the acoustic analogous parameters of the electrical model for the vocal tract, we can now derive the analogous sound pressure (the voltage in this model) wave as it travels along the dx length of the lossy tube (electrical line). The schematic diagram of functional components of the vocal tract along with a lossy electrical circuit model of every small length of the airways is shown in Fig. 2.1.

A Δx length of a lossy electrical line is illustrated in Fig. 2.1(b). Let x be the distance measured from the receiving end of the line, then $Z\Delta x$ is the series impedance of the Δx length of the line ($Z = R + jL\omega$) and $Y\Delta x$ is its shunt admittance ($Y = G + j c\omega$). The voltage at the end of Δx line is V and is the complex expression of the measured RMS voltage, whose magnitude and phase vary with distance along the line. As the line is lossy, the voltage at the other side of the Δx line is $V + g\Delta V$. By writing a KVL, we have

$$V + \Delta V = (I + \Delta I)Z\Delta x + Z\Delta x I + V \Rightarrow \frac{\Delta V}{\Delta x} = IZ + Z\Delta I.$$

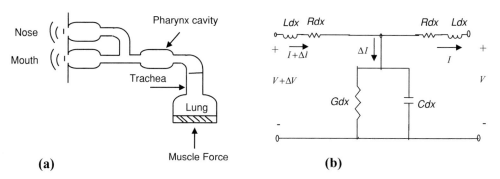

(a)   (b)

FIGURE 2.1: (a) Schematic diagram of functional components of the vocal tract; (b) electrical equivalent for a one-dimensional wave flowing through a lossy cylindrical pipe

As we let Δx approach zero, ΔV approaches dV and Δx approaches dx. The second term, which contains ΔI, can be neglected as it becomes a second-order differential equation and approaches zero much faster. Therefore at the limit it can be written as

$$\frac{dV}{dx} = IZ. \qquad (2.1)$$

Similarly by writing KCL and neglecting the second-order effects, we have

$$\frac{dI}{dx} = VY. \qquad (2.2)$$

By differentiating Eq. (2.1) and using Eq. (2.2), we obtain

$$\frac{d^2 V}{dx^2} = ZYV. \qquad (2.3)$$

The solution to Eq. (2.3) is

$$V = A_1 e^{(\sqrt{YZ})x} + A_2 e^{(-\sqrt{YZ})x}. \qquad (2.4)$$

Similarly, if we differentiate Eq. (2.1) and substitute Eq. (2.2) in it, we obtain

$$\frac{d^2 I}{dx^2} = ZYI. \qquad (2.5)$$

The solution to Eq. (2.5) is

$$I = B_1 e^{(\sqrt{YZ})x} + B_2 e^{(-\sqrt{YZ})x}. \qquad (2.6)$$

Constants A_1, A_2, B_1, and B_2 can be evaluated by using the conditions at the receiving end of the line when $x = 0$, $V = V_R$ and $I = I_R$. Substituting these values in Eqs. (4) and (2.6) yields

$$V = \frac{V_R + I_R Z_c}{2} e^{\gamma x} + \frac{V_R - I_R Z_c}{2} e^{-\gamma x}$$

$$I = \frac{V_R/Z_c + I_R}{2} e^{\gamma x} + \frac{V_R/Z_c - I_R}{2} e^{-\gamma x},$$

where $\gamma = \sqrt{ZY}$, which is called the propagation constant, and $Z_c = \sqrt{Z/Y}$, which is called the characteristic impedance of the line [1].

The electrical model discussed in this section represents the acoustic model mainly for the vocal tract. The acoustic model below the glottis has also been investigated by a number of researchers as briefly described below. Readers interested in the acoustic model for breath sound transmission based on the above electrical model may look at the references cited in the next section for further details.

2.2   RESPIRATORY SOUND GENERATION AND TRANSMISSION

The combination of the vocal tract and the subglottal airways including lungs form the respiratory tract, which has highly unique acoustic properties. The acoustic characteristics of the vocal tract and the subglottal airways have been modeled and investigated with the motivation to assess the relationship between the structure and the acoustic properties of the respiratory tract in healthy individuals and patients with respiratory disease [1–4]. To date, a number of acoustic models have been developed and investigated for respiratory sound transmission; however, there has not been a report indicating significant differences between the characteristics of the models for the two groups of healthy individuals and patients, which is mainly due to the fact that the models have not been applied to the patients' data in most of these studies. The main reason is probably that to model a biological organ one has to make many simplifications and hence reducing the sensitivity and specificity of the model to represent changes as a result of disease compared to the sensitivity of biological signals that can be recorded on the surface of the body and/or the clinical symptoms. Nevertheless, modeling a biological system can help better understand the mechanism; hence, indirectly helping the better diagnosis.

A common model for respiratory sound transmission is an electrical network of T-line circuits similar to that of the vocal tract. In the model described in [4] the acoustic properties of the respiratory tract were predicted and verified experimentally by modeling the respiratory tract as a cylindrical sound source entering a homogenous mixture of air bubbles in water with thermal losses, analogous to gas and fluid, that represented lung parenchyma. The model of parenchyma as a homogenous mixture of gas and fluid is justified considering the relatively low speed of sound in the parenchyma with respect to the free-field speed in either air or tissue. The speed of sound in such a mixture is about 2300 cm s^{-1} that is close to the sound propagation speed in trachea and the upper chest wall of humans [5]. The speed of sound through the parenchyma changes with the volume of the lung. It is at the maximum of 2500 cm s^{-1} in the deflated lung and decreases with a parabolic curve to the minimum of 2500 cm s^{-1} at total lung capacity [6, 7]. Since we normally breathe above functional residual capacity, FRC, the respiratory transmission models have been developed using the related values at FRC lung volume. The speed of sound at FRC is about 3500 cm s^{-1} [7]. Therefore, the sound wavelength at this speed for frequencies below 600 Hz is more than 5.8 cm. The assumption of respiratory sound transmission in most models is that the sound wavelengths of interest in the parenchyma are much longer than the alveolar radius. This assumption holds true for humans as the alveolar radius for an average adult is about 0.015 cm; hence much shorter than the sound wavelengths for frequencies below 600 Hz (5.8 cm).

Each airway segment is modeled by a T-equivalent electrical circuit similar to that of the vocal tract but with the addition of another shunt admittance to represent the acoustic properties of the airway walls (Fig. 2.2). A cascaded network of these T-circuits was used to

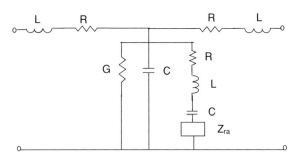

FIGURE 2.2: The T-lossy electrical circuit model representing a Δx length of each airway

represent a model of respiratory tract representing the vocal tract, trachea and the first five bronchial generations over the frequency range of 100 to 600 Hz [4]. This model proved to be adequate and provided a functional correlation between the sound speed and the density of lung parenchyma, which is dependent on the alveoli size. The sound speed increases when there are some collapsed alveoli as a result of respiratory diseases. This suggests that it might be possible to identify collapsed areas of the lungs by measuring the sound speed, which would provide a noninvasive diagnostic technique for monitoring lung diseases.

The above-mentioned model and other numerous studies either theoretically and experimentally have basically shown that an increase in the lung volume results in attenuation in the sound acceleration. The experimental studies are achieved by introducing a pseudorandom noise at the mouth of the human subject and recording the transmitted noise at different locations of the chest wall. A similar procedure has also been carried out on an isolated lung of a sheep, horse, and dog by introducing a noise to one side of the lung and recording the transmitted noise on the immediate opposite side of the lung under different gas volumes. While none of the sound transmission models explicitly predict attenuation at particular lung volumes, they predict a frequency-dependent increase in attenuation with the increasing gas–tissue ratio of the lung parenchyma. Thus, the larger amount of gas in the lungs at high lung volumes should lead to a greater attenuation. This has been supported by the experimental results reported in [8]. Theoretically, the speed of sound in a gas is inversely proportional to the square root of the mass of the gas and we know that the mass is equal to density multiplied by volume. Therefore, both density and volume can affect the sound speed.

A key question in this topic is how the sound is transmitted from the major airways to the chest wall. This issue has caused a considerable debate and discussions. In the model described above, it is assumed that all the sound is conducted to the chest wall by passing through the lung tissue. When the lung parenchyma is modeled as a homogenous mixture of gas bubbles in a liquid [4], the gas density should not play a role in attenuation of the sound in parenchyma. This has been supported by other studies that different gas densities have no significant effect on sound attenuation at least up to 400 Hz and most likely to 700 Hz [9, 10]. This finding suggests that the sound transmission occurs predominantly through lung tissue. Since it is not possible to study the effect of volume and density independently on sound transmission in human subjects, it may not be possible to exclude the possibility that changes in lung volume are responsible for the attenuation in sound transmission. Since the respiratory sound transmission is highly dispersive [7, 10–12], it seems that a change in lung volume should affect sound attenuation predominantly thorough associated changes in lung density.

CHAPTER 3

Breath Sounds Recording

Since the invention of stethoscope by the French physician, Laennec, in 1821, auscultation (listening to the sounds at body surface) has been the primary assessment technique for physicians. Despite the high cost of many modern stethoscopes, including digital stethoscopes, their use is limited to auscultation only as they are not usually tested, calibrated, or compared. Furthermore, they do not represent the full frequency spectrum of the sounds as they selectively amplify or attenuate sounds within the spectrum of clinical interest [13].

Digital data recording, on the other hand, provides a faithful representation of sounds. Fig. 3.1 shows the schematic of the most common respiratory sound recording. Respiratory

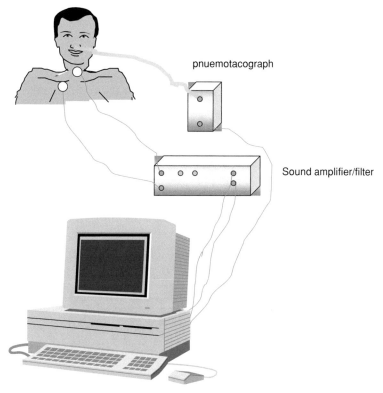

FIGURE 3.1: Typical apparatus for breath sounds recording

sounds are usually recorded either by electret microphones or sensitive contact accelerometers, amplified, filtered in the bandwidth of 50–2500 Hz and digitized by a sampling rate higher than at least 5 kHz. Respiratory flow is also commonly measured by a face mask or pnuemotachograph attached to a pressure transducer as shown in Fig. 3.1, and is digitized simultaneously with respiratory sounds. In fact, compared to other biological signals, the respiratory sound recording can be simpler as it can be recorded by a microphone, an audio preamplifier and a data acquisition (DAQ) card in place of which, as a start, one may even use the sound card of a computer. For research purposes, the recording apparatus must be chosen with more care though. The important factors are the noise level especially at low flow rates, the cut-off frequencies of the filter associated with the amplifier, the sensitivity of the sensor (specially if one uses accelerometers), the output voltage range of the amplifier to be matched with the input range of the DAQ, the input impedance of the amplifier as well as the sampling rate of the DAQ. In terms of the sensor to choose for recording respiratory sounds, there has been a long debate to choose accelerometers or microphones. However, as long as the frequency range of interest is below 5 kHz, there is not much difference in choosing either.

CHAPTER 4

Breath Sound Characteristics

Respiratory sounds have different characteristics depending on the location of recording. However, they are mainly divided into two classes: upper airway (tracheal) sounds usually recorded over the suprasternal notch of trachea, and lung sounds that are recorded over different locations of the chest wall either in the front or back. Tracheal sounds do not have much of diagnostic value as the upper airway may not be affected in serious lung diseases, while lung sounds have long been used for diagnosis purposes.

Lung sounds amplitude is different between persons and different locations on the chest surface and varies with flow. The peak of lung sound is in frequencies below 100 Hz. The lung sound energy drops off sharply between 100 and 200 Hz but it can still be detected at or above 800 Hz with sensitive microphones. The left top graph of Fig. 4.1 shows a typical airflow signal measured by a mouth-piece pneumotachograph. The positive values refer to inspiration and the negative values refer to expiration airflow. The left bottom graph shows the spectrogram (or sonogram) of the lung sound recorded simultaneously with that airflow signal. The spectrogram is a representation of the power spectrum for each time segment of the signal. The horizontal axis is the duration of the recording in seconds and the vertical axis is the frequency range. The magnitude of the power spectrum is therefore shown by color, where the pink color represents above 40 dB whereas the dark gray represents less than 4 dB of the power in Fig. 4.1. As it can be observed, the inspiration segments of the lung sound have much higher frequency components than expiration segments. In other words, inspiration sounds are louder than expiration sounds over the chest wall and this observation is fairly consistent among the subjects [14]. The right graph shows the average spectrum of all inspiration segments compared to that of expiration segments. Again, as it can be observed, there is about 6–10 dB difference between inspiration and expiration power spectra over a fairly large frequency range.

On the other hand, tracheal sound is strong and covers a wider frequency range than lung sound. Tracheal sound has a direct relationship with airflow and covers a frequency range up to 1500 Hz at the normal flow rate. Similar to the previous figure, the left graphs of Fig. 4.2 show a typical airflow signal on the top and the associated spectrogram of the tracheal signal on the bottom. As it can be observed, the tracheal sound signal is much louder than that of lung sound. However, the difference in inspiration and expiration power of the tracheal sound signal

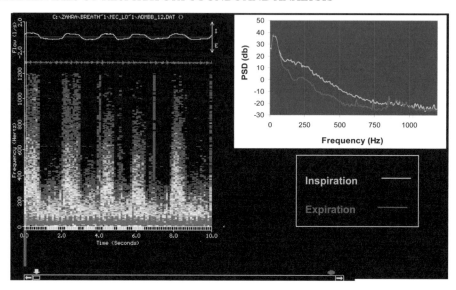

FIGURE 4.1: A typical lung sound signal spectrogram (left graph) along with the average spectra of inspiration and expiration (right graph) and the corresponding flow (top graph)

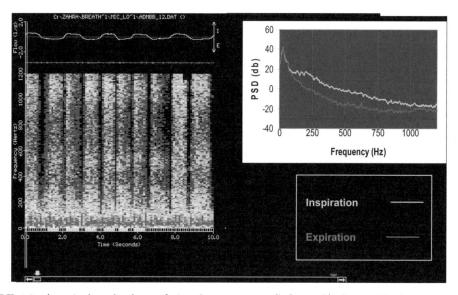

FIGURE 4.2: A typical tracheal sound signal spectrogram (left graph) along with the average spectra of inspiration and expiration (right graph) and the corresponding flow (top graph)

varies among the subjects greatly. In some people, there is not much difference while in others such as the subject in this example the expiratory sound is louder than the inspiratory sound.

The relationship of flow with power density of tracheal and lung sounds leads to the idea that at least the breath phases, i.e., inspiration/expiration, and the onset of breaths can be determined acoustically without the actual flow measurement; this was investigated a few years ago. The actual flow estimation by acoustical means, however, requires many more signal processing techniques and investigations. We will discuss this issue in more details in the following sections.

Like all other biological signals, respiratory sounds also differ among the subjects as their chest size and body mass are different. However, using digital signal processing techniques, researchers have sought methods to extract some characteristic features of the respiratory sounds that can be used for diagnostic purposes between healthy individuals and patients with various respiratory diseases. This has been the main motivation for most of respiratory sound researches, which we will address in more detail in the following sections.

CHAPTER 5

Current Research in Respiratory Acoustics

5.1 RESPIRATORY FLOW ESTIMATION

The relationship between respiratory sounds and flow has always been of great interest for researchers and physicians due to its diagnostic potentials. In a clinical respiratory and/or swallowing assessment, airflow is usually measured by spirometry devices, such as pneumotachograph, nasal cannulae connected to a pressure transducer, heated thermistor or anemometry. Airflow is also measured by indirect means, i.e., detection of chest and/or abdominal movements using respiratory inductance plethysmography (RIP), strain gauges, or magnetometers. The most reliable measurement of airflow is achieved by a mouth piece or face mask connected to a pneumotachograph [15]. However, this device cannot be used during the assessment of breathing and swallowing. Therefore, during the swallowing sound recording, airflow is usually measured by nasal cannulae connected to a pressure transducer. Potentially, this method could be an inaccurate measure of airflow because the air leaks around the nasal cannulae. In addition, if the subject's mouth breathes, the flow is not registered at all. For these reasons, the combined use of nasal cannulae connected to a pressure transducer and the measurement of respiratory inductance plethysmography (RIP) to monitor volume changes has been recommended as the best approach in assessing respiratory and swallowing patterns [15]. The application of these techniques has some disadvantages, especially when studying children with neurological impairments, in whom the study of swallowing is clinically important. Although the application of nasal cannulae may seem a minor intrusion, it can potentiate agitation in children with neurological impairments. In addition, applying the RIP devices is difficult in children with neurological impairments as their poor postural control and physical deformities can make it challenging to ensure stable positioning.

Due to the difficulties and inaccuracy of most of the flow measurement techniques as mentioned above, several researchers have attempted to estimate flow from respiratory sounds. Although many researchers studied the relationship between flow and respiratory sounds, few of them tried to actually estimate flow and address all its difficulties in real application [16–20].

As the first step in flow estimation by acoustical means, we also have to detect respiratory phases, i.e., inspiration and expiration, from the respiratory sounds. Our ear usually cannot distinguish the respiratory phases of the tracheal sound because the characteristics of inspiration and expiration sounds recorded at the trachea are very similar. On the other hand, lung sounds are significantly different between the two respiratory phases. In some locations of the chest wall, only the inspiration sounds can be heard. Therefore, the respiratory sounds recorded on the chest wall can be used as a signature for inspiratory sounds. This fact was used in a study [14] to develop an automated method that detected respiratory phases from simultaneously recorded tracheal and lung sounds. Since the method was based on the difference of the lung sound intensity between the two phases, lung sounds over different locations of the chest were investigated to find the location where the greatest difference in sound intensity between the inspiration and expiration phases was present. The results showed a minimum of 6 dB difference between the inspiratory and expiratory sounds within the 150–450 Hz range for the best recording location on the chest that yielded the greatest inspiration/expiration power difference.

Differences in sound intensity between the left and the right hemithorax can be heard by most listeners even in less than ideal test situations at approximately 2–3 dB difference. Auscultation, however, does not identify the location with the greatest difference between the respiratory phases on a specific side. The results of that study [14] showed that the left mid-clavicular area, second intercostals space (L1), and the right midclavicular area, third interspace (R3), are the common recording locations that yield the greatest power difference between the respiratory phases. Therefore, if the side of recording is determined by auscultation, then L1 or R3 locations can be chosen with confidence to record lung sound for respiratory phase detection.

The above-mentioned method uses the average power of the lung sounds over 150–450 Hz and a running window with the size of approximately half of a breath to detect the peaks of the calculated average power. The detected peaks are used as the signatures of the inspiration phases. Then, the local minima of the average power of the tracheal sounds are used to detect the onset of the breaths as the tracheal sounds are more sensitive to the variation in flow compared to the lung sounds; hence a more accurate breath onset can be obtained using tracheal sounds versus lung sounds. Respiratory phase detection by the suggested method has led to 100% accuracy as reported in [14]. The delay in breath onset detection was found to be in the range of 42 ms.

Given that the respiratory phases can be detected by the above-mentioned method, estimating the actual flow from the respiratory sounds has still been a challenge to meet. Tracheal sound mean amplitude, average power, mean power frequency, and the multiplication of tracheal sound mean frequency and mean amplitude in relation to flow were investigated. While the early studies [21] had suggested a power relationship between tracheal sound intensity and flow, recent studies have shown that the exponential model is superior to the polynomial and power models [17, 18].

Once the model to describe the flow–sound relationship is chosen, i.e., exponential model or power model, the flow can be estimated from the sound. However, all of the flow estimation models require calibration to tune the model parameters such that the model can overcome the high variation of the flow–sound relationship between different subjects. Therefore, all the flow estimation methods that have been developed to date have assumed that a few breaths of known flow of each subject are available for tuning the model. The dependence of the flow estimation on this calibration process is one of the great challenges to overcome in this field. Another issue that makes the flow estimation more challenging is that in most of the models suggested for flow estimation, once the model is tuned to estimate flow at one particular rate, i.e., tidal flow rate, the model is out of tune for a high or low flow rate estimation. This is because most of these models use tracheal average power in an exponential (or power) model for flow estimation but the parameters of the model differ for low, tidal, and high flow rates.

The need for calibration is the major drawback of flow estimation methods. Some of the above-mentioned methods achieved a reasonably low error in flow estimation, however all of them are heavily dependent on the calibration part; they need to have a copy of the breaths at every target flow during calibration. In one study [19] the flow rate was assumed to be constant and half of the data were used for calibration. In the other studies in which a variety of flow rates were considered, the model was calibrated with different sets of parameters at different flow rates assuming that a copy of every flow rate is available for calibration [17, 18, 22]. However, it is not always possible to capture respiratory sounds at different flow rates for calibration, especially when assessing young children, patients with neurological impairments, and/or patients in emergency conditions. Furthermore, the average error of these methods was more than 10% [16, 17, 19, 22], except in one [18] which was $5.8 \pm 3.0\%$ but at the cost of a much more complicated calibration procedure.

The respiratory sound features used in the above-mentioned studies to estimate flow are calculated from either the mean amplitude or average power of the sound signal. While these features in general can show a reasonable correspondence with flow at a particular flow rate, however, when they are used to estimate the flow at variable rates, they show a consistent undershoot or overshoot error. In other words, these features do not follow the target flow variation with one unique set of parameters of the model. To remedy this problem, these methods are in need of calibration for every target flow to tune the model to that target flow rate.

Respiratory sounds are stochastic signals and nonstationary in nature. Given the fact that the mean amplitude and average power are only the first- and second-order moments of the signal, they do not represent all statistical properties of respiratory sounds. In search of a feature of the respiratory sound that can follow flow variation, a recent study [20] presented a new method of flow estimation which used the entropy of the tracheal sound. Entropy is a measure of uncertainty of the signal and it involves calculation of probability density function (PDF) of the signal.

In that study [20], a modified linear model describing flow and the entropy of tracheal sound relationship was used for flow estimation at variable rates. The coefficients of the model were derived from only one breath sound sample with known flow at a medium flow rate (Fig. 5.1). Since heart sounds are an inevitable source of noise when recording lung sounds and its bandwidth has an overlap with the major components of the lung sounds, its effect has to be considered in flow estimation. It can be expected that the existence of the heart sounds in a portion of the lung sound record would change the PDF of that portion compared to the parts void of heart sounds. Hence, the presence of heart sounds may introduce an extra error in flow estimation especially at low flow rates.

A necessary step of the entropy-based method for flow estimation that is not shown in Fig. 5.1 is to remove the effect of heart sounds on the entropy of the lung sound record prior to deriving the model's coefficients. Therefore, heart sounds are first localized (with the method described in Section 5.3) in the calculated entropy signal and then that portion of the entropy signal is removed and interpolated to remove the heart sound effects [23].

The results of the entropy-based method showed that the model was able to follow the flow variation with a low error of about 9% [20]. The main advantage of the entropy-based

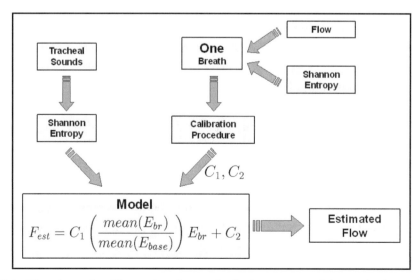

FIGURE 5.1: The schematic of flow estimation method using entropy of the tracheal sounds. Adopted from [20] with permission

model over all the previous studies is that it is robust in terms of flow variation, and that it does not need more than one breath with known flow at a tidal or medium flow rate for its calibration. Furthermore, with the aid of a new and simple technique to cancel the effect of heart sounds on the tracheal sound, the method was also able to estimate the flow at very shallow breathing, which had not been done in the previous studies. There have been major improvements over previous attempts in the flow estimation.

The above-mentioned methods are only a start for flow estimation. There are still several issues to be investigated. One is the need of calibration even though the last method described above has reduced this need significantly. There are some ideas under investigation whether this calibration process can be replaced by the use of a large data bank from many different normal subjects. The other issue to be considered is to evaluate the flow estimation methods in patients with various respiratory diseases. Will that still be a reliable method is something that has to be investigated. Yet another issue is to evaluate the method in noisy environments such as the emergency rooms.

5.2   HEART SOUND CANCELATION

As explained in the previous sections, lung sound analysis has been of interest for its diagnostic values for pathology of the lung and airways assessment. Most of the lung sounds energy is concentrated in the frequencies below 200 Hz, which has an overlap with the main frequency components of heart sounds. The heartbeat is an unavoidable source of interference for lung sound recording that when it occurs it changes both frequency and time characteristics of the lung sounds.

Physicians may try to ignore the heart sounds during auscultation or when assessing the recorded lung sounds remove the parts that include heart sounds. Since heart sounds occur regularly, the removal of heart sound included portions of the recorded sound signal causes artifacts and click sounds in those locations; hence making the signal unusable for any automatic analysis or even listening to the entire signal. Since both heart and lung sounds have major components in the frequency range below 200 Hz, a simple filtering cannot remove the effect of heart sounds. Hence, several researchers employed different methods to reduce or eliminate the effect of heart sounds in lung sounds recordings. The heart sound reduction methods can be divided in two groups: the methods that apply a filter to the entire lung sound record and reduce the heart sounds effect and the methods that remove the heart sound included portion of the lung sound record and then estimate the signal in the gaps. The recent methods developed for heart sound reduction are described below. Among these methods the wavelet denoising, adaptive filtering and independent component analysis belong to the first group of heart sound reduction methods, while the rest belong to the second group.

Wavelet denoising. Wavelet transform (WT) is a useful tool when dealing with signals that have some nonstationary parts. Therefore, some researchers have tried to develop filters based on wavelet transform. The WT-based adaptive filtering was first proposed in [24] to separate discontinuous adventitious sounds from vesicular sounds based on their nonstationary characteristics assuming that nonstationary parts of a signal in the time domain produce large WT coefficients (WTC) over many wavelet scales, whereas for the stationary parts the coefficients die out quickly with the increasing scale. Therefore, it is possible to apply a threshold to the WTC amplitudes to detect the most significant coefficients at each scale representing the nonstationary parts of the signal in the time domain; hence, the rest of the WTC correspond to stationary parts of the signal. Consequently, a wavelet domain separation of WTC corresponds to the time domain separation of stationary and nonstationary parts of the signal. This method was applied for heart sound separation (reduction) from lung sounds by a few researchers [24, 25]. The desired lung sounds and unwanted heart sounds portions of a signal were separated through iterative *multiresolution decomposition* and *multiresolution reconstruction* based on hard thresholding of the WTC. While the WT-based filters do reduce the heart sounds effect on the lung sound record, however, according to the reported results, some audible heart sounds still remained in the lung sound record.

Adaptive filtering. Linear adaptive filtering (usually with the least mean squares (LMS) or recursive least squares (RLS) algorithms for adaptation) has been used widely for canceling power-line noise in biological signals successfully. The technique is to use the original signal including the noise as one of the inputs and a copy of the noise signal as another input in which the filter tries to find the components in the main signal that match with the copy of the noise and finally after the filter coefficients are optimized by an algorithm, i.e., LMS or RLS, the adapted-to-noise output is subtracted from the signal; hence the remaining is supposedly the noise-free signal.

This technique works well if the signal and noise are uncorrelated as it is an assumption in the design of linear adaptive filters. This assumption holds true for the power-line noise and biological signals. However, in applying this technique to separate heart and lung sounds from each other, the main problem is how to provide a copy of the noise, i.e., heart sounds in this case, that is uncorrelated to the main signal, i.e., lung sounds. Even if we assume that the heart and lung sounds are uncorrelated in their source of generation, they become correlated when we record them on the surface of the chest wall as they both pass through the same medium.

Two groups, who applied adaptive filtering for heart sound reduction, used ECG signals instead of the heart sound as the copy of the noise input [26, 27]. This is highly questionable as the ECG signal is not the noise of the lung sound record and hence even if the filter tries to adapt itself to the ECG signal, the results will not be meaningful. Two other groups used the heart sounds recorded over the heart of the subjects as the reference signal for the adaptive filtering with the LMS algorithm [28, 29]. However, the heart sound signal recorded over the chest wall inevitably includes lung sounds as well.

In order to eliminate the need of the reference signal for the adaptive filter, some researchers proposed a single recording technique based on the modified version of the adaptive LMS algorithm by adding a low-pass filter with a cut-off frequency of 250 Hz in the error signal path to the filter [30]. However, their results showed that heart sounds were still clearly audible due to the improper identification of the heart sound segments within the long sound record.

As it can be deduced from the above review, the reference signal plays a key role in the successful noise reduction in the output of the linear adaptive filter. In one of the successful and most recent studies using adaptive filtering, the reference signal was constructed by first detecting the heart sounds locations in the bandpass filtered version of the original record using an old method introduced in [31] and then replacing all the components of the bandpass filtered original signal by zero except for the heart sound included parts. The resulted impulse-like signal was used as the reference signal for the adaptive noise cancelation using the RLS algorithm. The results were evaluated both quantitatively and qualitatively. The average power spectral densities of unfiltered signals over four frequency bands in areas including and excluding heart sounds were compared with those of the filtered recordings. Also a panel of researchers experienced in lung sound auscultation provided qualitative analyses of the filtered signals, which was favorable for significant heart sound reduction [32].

Spectrogram-independent component analysis. Blind source separation (BSS) is a method to recover independent source signals while only a linear mixture of the signals is available, in which the sources and the way they have been mixed is unknown; hence the term "blind" in the method. Independent Component Analysis (ICA) is a technique used to solve BSS problems. ICA finds a linear coordinate system such that the recovered signals are statistically independent [33].

However, in some cases the signals are not linearly mixed but are convolved due to transmission through one or more mediums with unknown transfer functions. Therefore only mixed and distorted versions of the signals are available. Applying ICA on the spectrograms of the mixed signals has been suggested as an alternative solution to this kind of problem [33].

In the case of lung and heart sounds, since both sounds pass through fat, skin and lung tissues, it makes sense that the signals are a convolutive mixtures. This technique has been applied for separating speech signals [34] and for heart sound reduction [35]. The assumption in applying this method is that the source signals are originally statistically independent and at least one of them has the Gaussian distribution. At least two versions of the mixed signals are required although the more the better the performance of separations, which is also at a much higher computational cost.

In [35] lung sounds were recorded from two sensors on the chest wall to provide two mixed signals, and then ICA was applied on the spectrograms of the signals. In general, it is difficult to evaluate the results of ICA as the method's claim is that it recovers the original signals at the source while there is no access to the actual source signals to compare with and evaluate. Therefore, in heart sound cancelation we can only compare it quantitatively using the spectra of the recorded lung sound portions free of heart sound and the resulted signal for any significant changes and also by visual and auditory means. The results of the ICA-based method have been shown to reduce the effect of heart sound while still a weakened heart sound could be heard in the resultant signal. In addition, the lung sounds were heard somewhat different from the original signal [35]. This is expected because the techniques that are applied to the entire signal inevitably change the lung sounds portions free of heart sound too.

Time–frequency filtering. The idea of removing the heart sound portions of the lung sound record and then estimating the missing data was first introduced in [36], in which the heart sound segments are removed from the lung sound record using time-varying filtering based on the short-time Fourier transform (STFT) of the original signal and then interpolation is used to fill the gaps in the time–frequency plane. Finally, the signal is reconstructed in the time domain. The block diagram in Fig. 5.2 illustrates the necessary steps of this method. The performance of the method is also shown by an example in Fig. 5.3. The result of this method by far has been the best among the other heart sound cancelation methods. It is also computationally efficient.

Multiscale product wavelet filtering and linear prediction. In this method [38, 39] the heart sound segments are detected by multiscale product (described in the next section) and removed from the wavelet coefficients at every scale. After removal of the heart sound included segments, the missing data are estimated by the linear prediction using either auto regressive (AR) or moving average (MA) models. AR and MA models are two common signal processing tools used to predict past or future values of a time-limited signal. The predicted samples are basically weighted linear combinations of the signal known values. A more complete discussion about AR/MA models can be found in [40].

This idea has been used for interpolation of missing speech segments in audio signals for short periods of time that can be assumed stationary [41]. However, it is too simplistic to assume that lung sounds are stationary during the entire duration of a respiratory cycle (inspiration/expiration) especially in the vicinity of breath onsets. Therefore, correct selection

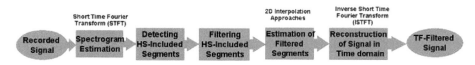

FIGURE 5.2: Block diagram of the STFT method. Adopted from [37] with permission

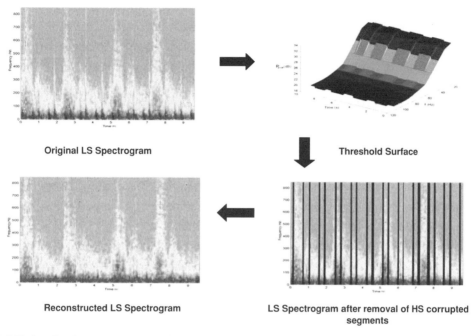

Original LS Spectrogram

Threshold Surface

Reconstructed LS Spectrogram

LS Spectrogram after removal of HS corrupted segments

FIGURE 5.3: Performance of the STFT method for a typical lung sound record

of the order as well as the type of the linear prediction model (AR or MA) must be done carefully to ensure that the data used for prediction of the gaps are indeed stationary.

The choice of either AR or MA modeling depends on the location of the heart sound segment within the breath cycle. If heart sound occurs close to the vicinity of the onset of the respiratory phase, i.e., inspiration or expiration, the method uses MA modeling to predict the gap with future values, due to the fact that most of the information about the samples in the gap is in the next lung sound segment; otherwise the method uses AR modeling.

The order of either AR or MA model is increased until the energy difference between the original and estimated data is minimum. The order, where the performance of the AR or MA models results in the minimum energy difference between the original and extrapolated data, is chosen as the best order for that portion of the lung sound signal. The performance of this method is comparable with the time–frequency filtering method both quantitatively and qualitatively, but this method is more complex.

Among the heart sound cancelation methods, those that apply a filter to the entire record of lung sound do reduce the heart sound but do not remove their effect completely, while the lung sound also changes slightly as the filter is applied to the entire signal. The time–frequency filtering method, multiscale wavelet filtering method, and linear prediction method perform better compared to the other methods because they remove the entire heart sound included segments and estimate the missing data by either spline interpolation or linear prediction. Therefore, heart sounds are completely removed from the record and the rest of the data remain unchanged. Removal of the heart sound included segments followed by interpolation or linear prediction for the missing data makes the record to be heard unchanged but without heart sounds.

Recurrence time and nonlinear prediction. In this method [42], the heart sound locations are found by the recurrence time statistic measure (described in Section 5.3), and removed from the original lung sound record like in the previous methods in this group. The missing data are then predicted by a nonlinear scheme. In fact, since the heart sound localization is achieved in state space, the prediction is also in the state space using the six neighboring trajectories to reconstruct the punctured space after removing heart sound portions. Then, the predicted reconstructed data are brought back into the time domain.

The results of this method seem to be reasonable as the reported differences in the spectra of the original lung sound portions free of heart sounds and the predicted lung sounds are less than 3 dB in the frequency ranges below 300 Hz. Since our ear is not sensitive to the differences below 3 dB, it can be concluded that the predicted data with this method are not heard much different from the original lung sounds free of heart sounds. However, the study does not elaborate on the differences at different flow rates and whether the method is successful in high flow rates.

5.3   HEART SOUND LOCALIZATION

The first step in all heart sounds reduction techniques except the wavelet denoising and ICA analysis is to localize the heart sound included segments in a lung sound record. Several methods have been proposed for heart sound localization. In [23, 30, 36, 38, 43] an adaptive thresholding on lung sound signal either in the time or time–frequency domains within a moving window was used to detect heart sounds included segments. Fixed-thresholding the variance fractal dimension trajectory (VFDT) of the lung sounds in a moving window was also suggested in [44]. In general, a threshold is applied to a feature extracted from the lung sounds record. The extracted features of the lung sound signal to be thresholded in the above-mentioned methods are average power, variance of the signal in the time domain, average power of the fifth-level wavelet coefficients [36], multiscale product of the wavelet coefficients [38], VFDT of the lung sounds [44], and entropy (using the Shanon entropy equation) of the lung sounds [23]. Yet in another study [45] heart sounds locations were detected using an adaptive recursive least-squares method, in which the signal was matched with a delayed version of itself and the error of the adaptive filter was thresholded. Also recurrence time statistics was used in another study to detect heart sounds [42]. In order to discuss the advantages and disadvantages of these methods, each is briefly described below followed by a summary of comparison between the methods.

Average power based method. The spectra of lung sound segments including heart sounds usually have higher intensity in the range of 20–300 Hz specially at low and medium flow rates. Therefore, in this method, the spectrogram of the signal is analyzed as an image with M gray levels and the locations of heart sound segments are detected by using an adaptive threshold. In order to specify such a threshold, the average power of the inspiration and expiration segments of the original recorded lung sound signal is calculated over the frequency band of 20–40 Hz to define two reference thresholds (Th) [36],

$$\text{Th} = \mu_{P_{\text{ave}}} + k_{\text{adj}} \times \sigma_{P_{\text{ave}}},$$

where $\mu_{P_{\text{ave}}}$ and $\sigma_{P_{\text{ave}}}$ are the mean value and the standard deviation of the calculated average power of the inspiration or expiration segments and k_{adj} is a threshold adjustment parameter that needs to be determined at the training stage by selecting some known segments with and void of heart sounds. Then, these two thresholds are applied to the spectrogram of the recorded lung sound signal. The average power of every segment within 20–40 Hz is compared with the threshold. The segment with one or more points above the threshold is considered as a heart sound included segment. The total reported error of this method in heart sound detection has been about 7% (details in Tables 5.1 and 5.2). This method is fast and simple to implement but requires training to tune it for every subject. Therefore, it cannot be run in a fully automated mode.

Wavelet coefficient based method. This method is very similar to the previous one but with two differences. Instead of the spectrogram of the original lung sound signal, the spectrogram of the fifth-level coefficient of the wavelet transform of the original signal is calculated. Also, the threshold in this method does not need any tuning and is simply the mean of the average power of the original lung sound signal. The error of this method is reported to be about 10%

TABLE 5.1: Mean and Standard Deviation Error (%) of Different Heart Sounds Localization Methods at a Low Flow Rate. Adopted from [23] with Permission

METHOD	ERROR (%)		
	FALSE NEGATIVE	FALSE POSITIVE	OVERALL
Entropy	0.0 ± 0.0	0.1 ± 0.4	0.1 ± 0.4
Multiscale product	0.8 ± 1.1	0.6 ± 0.9	1.4 ± 1.1
Average power	2.8 ± 1.4	3.3 ± 1.8	6.1 ± 2.0
Variance	1.9 ± 1.8	7.0 ± 1.6	9.0 ± 2.0
Fifth wavelet coefficient	1.1 ± 1.1	6.8 ± 6.3	7.9 ± 6.5

TABLE 5.2: Mean and Standard Deviation Error (%) of Different Heart Sounds Localization Methods at a Medium Flow Rate. Partially Adopted from [23] with Permission

METHOD	ERROR (%)		
	FALSE NEGATIVE	FALSE POSITIVE	OVERALL
Entropy	1.0 ± 0.7	0.0 ± 0.0	1.0 ± 0.7
Multiresolution product	2.2 ± 3.0	1.2 ± 1.0	3.4 ± 2.2
Average power	3.6 ± 2.2	3.7 ± 2.4	7.3 ± 2.0
Variance	2.0 ± 1.6	6.9 ± 2.0	8.9 ± 3.3
Fifth wavelet coefficient	0.7 ± 0.7	10.2 ± 5.6	10.9 ± 6.0
Recurrence time	8%	4%	NA

but with a very low false negative error (1%) [46]. This method can be used in a fully automated mode with a reasonable speed.

Variance-based method. This method is the simplest compared to the other methods, in which the variance of the original lung sound signal is calculated in a 100 ms moving window and then a threshold that is simply the average of the variances of all the windows is applied and the segments above the threshold are considered as the heart sound included segments. The error of this method is reported to be about 9% with a low false negative error (less than 2%) [46]. The method is fast, fully automated, and easy to be implemented.

Variance fractal dimension based method. Variance fractal dimension (VFD) is a measure of the signal's complexity in terms of morphology. VFD is defined based on the power-law relationship between the variance of the amplitude increments of a signal, $B(t)$, which is produced by a dynamical process (and modeled as fractional Brownian motion for the VFD analysis) over a time increment $\Delta t = |t_2 - t_1|$, with $B(t_2) - B(t_1)$ denoted as $(\Delta B)_{\Delta t}$. $B(t)$ is the signal of interest that in this case is the lung sound signal. The power law is as follows,

$$\mathrm{Var}[(\Delta B)_{\Delta t}] \sim \Delta t^{2H},$$

where H is the Hurst exponent:

$$H = \lim_{\Delta t \to 0} \left[\frac{1}{2} \frac{\log_b[\mathrm{Var}(\Delta B)_{\Delta t}]}{\log_b(\Delta t)} \right].$$

Then the VFD for a process with embedded Euclidean dimension, E, is determined by [47]

$$D_\sigma = E + 1 - H.$$

In the VFD-based method for heart sound localization, the VFD of the original lung sound signal is calculated in a moving window of 100 ms with an increment size of 25 ms. Then, the resulted VFD trajectory is compared with an adaptive threshold that is tuned for every subject, and the segments above the threshold are considered as the heart sound included segments. The accuracy of this method is reported to be in the range of 61−96% for the first and second heart sounds for different flow rates [44]. This method is slow in computation and its accuracy is very sensitive to the window size of VFD calculation. It also requires tuning the threshold for every subject.

Multiscale product based method. This method is based on the properties of the product of wavelet coefficients at several scales. The different behavior of signal and noise in the wavelet domain can be analyzed using the concept of Lipschitz regularity. As an example, the Lipschitz regularity of a step function is 0. If the function is smoother than a step, then its Lipschitz regularity is positive. On the other hand, the Lipschitz regularity of the delta-function is −1, while a Gaussian noise generally has a Lipschitz regularity of −1/2 [41]. A function $f(n)$ has uniformly Lipschitz regularity value α between 0 and 1 over an interval $[a, b]$ if and only if there exists a constant $K > 0$ such that for all $n \in [a, b]$, the wavelet transform satisfies,

$$\left| W_j f(n) \right| \leq K(2^j)^\alpha,$$

where j is the jth scale level in the wavelet domain.

The above equation implies that the magnitude of the singularities, i.e., the portions of the signal that are not of background Gaussian noise and are singular in nature such as the heart sound within lung sound record, increases along the scales. On the other hand, the wavelet transform magnitude of the signals with negative Lipschitz regularity, i.e., Gaussian noise, decreases as the scale increases. Thus, the multiplication of the wavelet coefficients between the decomposition levels can lead to identification of singularities. This is a very desirable property that has been used in [48] to enhance image borders and reduce noise, and in [49] to reduce clutter in radar signals and also in [38] for heart sound localization.

The method introduced in [38] decomposes the original signal into three levels in the wavelet domain using the Symlet (order 5) wavelet and the product of the wavelet coefficients is calculated at level 3. Then, a threshold which is simply the mean plus five times the standard deviation of the heart sound free portions of the original lung sounds is applied to the multiscale product and the coefficients above the threshold are found to be at the locations of the heart sounds. The error of this method was reported in the range of 1−3% but its false positive error was slightly higher than false negative error. The method is semiautomated for determining its threshold.

Recurrence time statistics based method. This method is based on calculating the recurrence time statistics of the lung sound record [42]. It is the application of the nonlinear dynamic metric tools for reconstructing the signal in a larger dimension state space and then measuring the number of the recurrence of the states above a threshold. The recurrence time statistic is determined as the number of the states in the similarity set (calculated in a running window over the entire original signal), in which the signal is close to a predefined reference state. The peaks of this measure of recurrence time statistic presumably correspond to the heart sound locations and hence can be detected by a threshold. Once the lung sound record was reconstructed with an embedding dimension of 7, it was observed that the attractors of lung sound portions including and void of heart sound were significantly different [42].

This method has resulted in 4% false positive and 8% false negative error in heart sound detection. It is also reported to be quite sensitive to the value of closeness to determine the recurrence time statistic measure [42].

Entropy-based method. This method [23] calculates the entropy of the lung sound record using Shanon entropy. Entropy is usually known as a measure of uncertainty of a process and it involves calculation of the probability density function (PDF). Therefore, it can be expected to reveal more of the statistical information about the signal than, for instance, the average power. As shown in Fig. 5.4 the PDF of the lung sound segments free of heart sounds is concentrated around zero, while the PDF of the heart sounds included segments is expanded over a wider range. Thus, the entropy of the heart sounds included segments is greater than that of the segments free of heart sounds, and can be used as a mean for heart sounds localization.

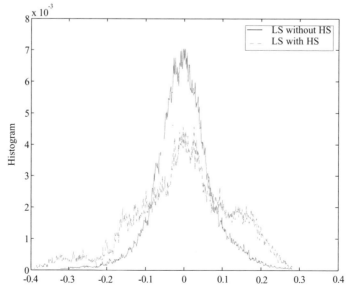

FIGURE 5.4: Histogram of the lung sound segments including heart sounds (solid curve) and void of heart sounds (dashed curve); Adopted from [23] with permission

For a set of events with PDF of $\{P_i, i=1, \ldots, N\}$ the Shannon entropy is defined as [50]

$$H(p) = -\sum_{i=1}^{N} P_i \log P_i.$$

For events with probability around 0 or 1, $P_i \log(P_i)$converges to zero and the entropy is minimum, while for random signals with a uniform PDF such as pure noise, the entropy is maximum. Since entropy is based on the PDF of the signal, accurate estimation of the PDF is of great importance in calculating entropy. Although histogram is an easy and fast way for estimating the PDF, its accuracy declines when the number of samples is low. Therefore, in [23], the PDF was calculated using the normal kernel function. If N independent observations, $\{X_1, \ldots, X_N\}$, of the random variable, x, with probability density function $p(x)$is given, then its kernel estimator $\hat{P}_i(x)$is defined as

$$\hat{P}_i(x) = \frac{1}{N} \sum_{i=1}^{N} \frac{1}{h} K\left(\frac{x - X_i}{h}\right),$$

where $K(x)$ is the kernel function and h is its bandwidth. The normal kernel function is defined as

$$K(x) = \frac{1}{\sqrt{2\pi}} e^{-x^2/2}.$$

The optimum value of h for the normal kernel to reduce the bias and variance of the estimator is approximated as [51]

$$h_{\text{opt}} = 1.06\hat{\sigma}(x) N^{-0.2},$$

where $\hat{\sigma}(x)$is the standard deviation of the input observations and $N$is the number of observation samples. The interested reader is referred to [23] for detailed information about PDF estimation and entropy calculation.

The entropy of lung sound records in this method is calculated in windows of 20 ms with a 50% overlap between the successive windows. Mean plus standard deviation $(\mu + \sigma)$ of the calculated entropy is used as the threshold to detect heart sound locations, i.e., the values above this threshold correspond to the heart sound locations in the original signal. This method has been reported to have a very low error (about 1%) in heart sound detection with a high resolution in boundary locations of the heart sound. The method is fully automated and robust to the variations between the subjects.

Comparison Between the Heart Sound Localization Methods

Since heart sound localization was used as part of the heart sound cancelation method, some of the above studies have not reported the localization errors. Also not all the studies have used the same protocol for data collection. While the amount of target flow affects the accuracy of the heart sound detection, this issue has not been addressed in some studies; making the comparison difficult. Tables 5.1 and 5.2 show the reported errors of the above-mentioned studies.

Other than the accuracy in heart sound localization, one should be careful about the false positive and negative errors of the method as well. As heart sounds localization is a preprocessing step for most of the heart sounds reduction methods, the techniques with smaller false negative error are preferred. In other words, it is more important not to miss any heart sound included segment than detecting some segments free of heart sound as the segments including heart sounds. Therefore, if the overall errors of some methods are comparable, the method with smaller false negative error is preferred. Furthermore, since all of the heart sound localization methods eventually threshold an extracted feature of the lung sound record to detect heart sound portions, the methods can be compared based on their robustness and dependence on the window size or overlap between the windows when calculating the feature of the lung sound that is to be thresholded for heart sound localization. In addition, some of the above-mentioned methods are automated and some require tuning the threshold for every subject's data. Hence, the mode of the method can also be an issue of interest. The computational cost of the method is another issue that may become of consideration especially for on-line applications.

The method based on VFDT is a semiautomated method and highly sensitive to the window size and overlap between the successive windows for VFDT calculation and the overall error was about 5–15% [44]. The methods of average power, variance, and fifth-level wavelet coefficient were all calculated in windows of 100 ms with a 50% overlap between the successive windows. Therefore the resolution of these methods for the boundary location of the heart sound segments is limited to 100 ms. In the entropy-based method the window size was chosen as 20 ms as this size was found to be the optimum window size in terms of the overall accuracy of the method; hence, this method has a higher resolution than the others. The multiresolution product based method does not use any fixed window size. Hence, theoretically one may expect its resolution to be higher than that of the other methods. However, due to its higher false positive error (Tables 5.1 and 5.2) and merging the small segments in the algorithm its heart sound included segments' width is greater than that of the entropy-based method. On the other hand, the multiresolution product based method is superior to the entropy-based method in terms of computational cost and speed.

As the results in Tables 5.1 and 5.2 indicate the performance of the Shannon entropy based method is superior to all other methods in terms of accuracy. The total average error and its standard deviation are significantly lower than those of the other methods. Furthermore, this method is fully automated without any additional need for adjusting the method to respiratory phases or subjects. The second best method for heart sounds localization is multiresolution product, which relies on the nonstationary characteristics of lung sound records in the presence

TABLE 5.3: Comparison of Different Methods for Heart Sounds Localization. Adopted Partially from [23] with Permission

METHOD	ERROR			RESOLUTION	SPEED	SENSITIVITY TO WINDOW SIZE	MODE
	FALSE NEGATIVE	OVERALL	STANDARD DEVIATION				
Entropy	Low	Low	Low	High	Low	Low	Automatic
Multiscale product	Low	Medium	Medium	Medium	Medium	Low	Semiautomatic
Average power	Medium	High	Medium	Low	Medium	Medium	Semiautomatic
Variance	Low	High	Medium	Low	High	Low	Automatic
5th Wavelet coefficient	Low	High	High	Low	Medium	Medium	Automatic
VFDT	-	High	High	Low	Low	High	Semiautomatic
Recurrence Time	High	Medium	-	Low	Low	High	Semiautomatic

of heart sounds. However, not only heart sound included segments of the lung sounds are highly nonstationary, but also lung sounds are nonstationary signals in general. It can be shown that lung sounds are stationary only in the upper 40–20% values of the target flow, where the flow plateaus. Therefore, the accuracy of the heart sounds localization method using multiresolution product declines when heart sounds occur around the onset of the breath, where lung sound segments void of heart sounds are also nonstationary. For this reason, the multiresolution product method has a higher false positive error compared to that of the entropy-based method.

Table 5.3 shows a summary comparison of the different heart sound localization methods discussed in this section. The methods are compared in terms of false negative, standard deviation and overall values of errors, resolution, speed, sensitivity to window size and operational mode. The values of error were categorized in three groups of low (<3%), medium (3% < error <6%), and high (>6%). Also standard deviation of the error values was categorized as low (<1%), medium (<3%), and high (>3%).

CHAPTER 6

Nonlinear Analysis of Lung Sounds for Diagnostic Purposes

Diagnosis of respiratory diseases in the absence of the adventitious sounds, i.e., wheezes, crackles, etc., is more challenging without the use of conventional methods such as chest x-ray, MRI, and CT that are invasive methods. However, some adventitious sounds such as wheezes are found to have a very low sensitivity to be used as a reliable indicator of airway narrowing in clinical practice [52]. On the other hand, lung sound intensity decrease is shown to be more consistently associated with bronchial narrowing after methacholine challenge [53, 54].

Adventitious sounds are relatively easily detectable both subjectively by auscultation and objectively by running some simple signal processing techniques. However, as mentioned above, adventitious sounds are not always present in all respiratory diseases and subtle changes in lung sounds may not be revealed by traditional spectral analysis. Therefore, in recent years researchers have attempted to apply nonlinear analysis to lung sounds for extracting diagnostic signatures. Waveform fractal dimension and state space parameters have been used for analysis of lung sounds in [54, 55], and also have been applied to swallowing [56] and speech [57] sounds with encouraging results. Hence, in this section these analyses and their results on lung sounds are described briefly.

Waveform fractal dimension (WFD). This feature is a morphological measure of the complexity of the signal. There are several algorithms to calculate fractal dimensions. One algorithm which seems to be more robust in terms of window size and also showing more sensitivity to lung sounds changes [58] is the algorithm presented by Katz in 1988 [59],

$$\text{WFD} = \frac{\log_{10}(n)}{\log_{10}(n) + \log_{10}(d/L)},$$

where n is the number of increments between samples of the waveform over which KFD is calculated, L is the sum of all of the distances between successive increments, and d is the value of the maximum distance measured from the beginning of the first increment.

In [54] the lung sound records before and after methacholine challenge in eight children suspicious of bronchoconstriction were obtained. Subjects were grouped as group 1 whose forceful expiratory volume (FEV) changed more than 20%, and group 2 whose FEV drop was less than 20% after the final step of methacholine challenge but might have still exhibited mild bronchoconstriction, but were not considered as having hyperreactive airways. The study used the RMS value as well as KFD values of the lung sound intensity in order to classify the signals pre- and postmethacholine challenge. The results showed that adding the KFD features to the classification increased the accuracy significantly. The accuracies of the combined KFD and RMS features for classifying the signals into pre- and postmethacholine challenge were about 90% and 80% for the two groups, respectively.

State space parameters. Embedding dimension and time-delay values are necessary to reconstruct a signal's attractor in state space. Using the Takens method of delays [60], the E-dimensional vectors $X(k)$ can be constructed as

$$X(k) = [x(k), x(k+\tau), x(k+2\tau), \ldots, x(k+(E-1)\tau)], \quad k = 1, 2, \ldots, N-(E-1)\tau,$$

where $X(k)$ is one point of the trajectory in the phase space at time k, τ is an appropriate time lag (an integer multiple of the sampling period), and E is called the embedding dimension and determines the smallest number of independent variables that uniquely describe the character of the system.

Choosing the appropriate time lag, τ, is a challenge because if it is too small, it does not provide enough distinct information between the successive vectors and result in the attractor being very close to the diagonal or identity line of the embedding space. On the other hand, if τ is very large, the successive vectors may become casually unrelated and the attractor will appear to wander all over the state space. The right selection of τ will become more important when working with finite length experimental time series and/or noisy data sets [61]. An approximation for τ may be obtained by the lag at which the autocorrelation of the time series drops to $(1-1/e)$ of its maximum value [60] or by the geometry-based method introduced in [61] that measures the average displacement of the embedding vectors from their original locations on the line of identity and chooses the time delay where the slope of the curve of average displacement versus time delay decreases to less than 40% of its initial value. The purpose of time-delay embedding is to unfold the attractor such that all self-crossings of the orbit can be eliminated. The attractor will be unfolded if we use any dimension greater or equal to the minimum embedding dimension $E_{\min}$. The minimum embedding dimension is usually determined by the method of false nearest neighbor [60]. To determine the minimum embedding dimension, it is increased until no false neighbor is remained.

Lyapunov exponents.   These parameters are metric tools commonly used to detect the chaotic nature of dynamic but deterministic systems. Lyapunov exponents were originally developed in classical mechanics to examine a system's stability [62]. Lyapunov exponents are measures of the exponential rate of divergence or convergence of nearby state space trajectories and the sign of exponents, i.e., positive or negative, provides indication of divergence or convergence (respectively) of a system's energy [57, 63, 64]. Therefore, if a system has positive exponents, with the smallest change in initial conditions the trajectory quickly follows a completely different path. The more positive the exponents, the faster the nearby trajectories move apart. Therefore, if a system is known to be deterministic, a positive Lyapunov can be taken as the proof of the system's sensitivity to the initial condition and presence of chaos. If both positive and negative exponents exist, two different initial conditions will diverge exponentially at a rate given by the largest Lyapunov exponent [57, 61, 63, 64].

State space parameters as mentioned above have been determined for lung sounds in six healthy subjects and one patient at pre- and postmethacholine challenge at low, medium, and high flow rates in [65]. The results showed that the embedding dimension of the lung sounds in healthy subjects was consistently around 9 regardless of the flow rate. The patient's embedding dimension after methacholine challenge decreased noticeably. The time delays significantly decreased when it was determined for the filtered lung sound in the frequency range of 150–450 Hz. This has been probably due to the reduced effect of heart sound in that frequency range. However, it has to be studied further. Another state space parameter that showed changes to the reduction of heart sound and also a very significant change to the methacholine challenge was the number of positive Lyapunov exponents. Overall, it decreased for the filtered lung sounds. While the patient's number of positive Lyapunov exponents was similar to that of healthy subjects before methacholine challenge, it decreased significantly after (from 92% to 25%).

State space parameters were also calculated in [42] but with the purpose of heart sound localization and cancelation as explained in Section 5.3. The minimum embedding dimension found in [42] was 7, which is lower than what was found in [65]. The time delays found in the later study were similar to those found in the former but similar to the values calculated on the filtered signals. However, from the second study [42] it is not clear whether the time-delay values have been found on the original signal or on the portions of the lung sound free of heart sounds. In the later case, this result is congruent with the speculation in [65] that the much larger time delays of the unfiltered lung sounds are probably due to the presence of the heart sounds.

In another study [66], the chaotic nature of lung sounds was investigated by calculating the Lyapunov spectra and correlation dimension; both indicated the presence of chaos in lung sounds. The sum of the Lyapunov exponents was found to be negative but the largest exponent was found positive in all of the 16 test subjects.

CHAPTER 7

Adventitious Sound Detection

7.1 COMMON SYMPTOMATIC LUNG SOUNDS

The common symptomatic lung sounds are bronchial, fine and coarse crackles, wheezes, squawks, rhonchi, pleural rub, and stridor sounds. Each of these sounds is described briefly below. A thorough discussion about the origin of these sounds and associations with particular diseases can be found in [53, 67].

Bronchial sounds. This sound is similar to the normal breathing heard over large airways; however, if heard over the chest, away from large airways, it is considered as an abnormal sound. It is often referred to as a tubular sound as it sounds like blowing through a tube. Fig. 7.1 shows a bronchial sound in the time and time–frequency domain. Recall that for normal lung sounds, the inspiration is much louder than expiration. However, a bronchial sound is usually the opposite and for this reason it is abnormal when heard over the chest wall.

Crackle sounds. Crackles are discontinuous or explosive sounds that are usually caused by airway opening and airway secretion. They are usually a sign of disease. Crackles are divided into two groups of *fine* and *coarse* crackles. Figs. 7.2 and 7.3 show examples of fine versus coarse crackles in the time and time–frequency domains. In general, as it can also be observed in the figures, fine crackles are higher frequency sounds compared to coarse crackles. Coarse crackles have less intensity and are longer in duration. For this reason, they are also called "explosive" and "bubbling" sounds to describe fine and coarse crackles, respectively. Crackles are present in heart congestion failure, pneumonia, bronchiectasis, pulmonary fibrosis, or chronic diffuse parenchymal lung disease. The location and number of heard crackles are some indicators of the type of disease.

Wheezes. The high-pitched whistling music type of sounds heard over large airways as well as over the chest are called wheezes. Fig. 7.4 shows a typical wheezing sound in the time and time–frequency domain. Wheezes can be caused by airway narrowing and the increased secretions. Wheezes are usually heard in congestive heart failure, asthma, pneumonia, chronic bronchitis and emphysema, bronchiectasis.

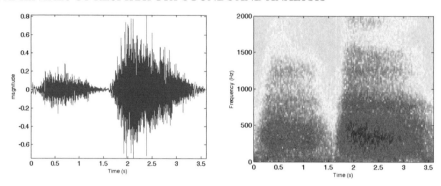

FIGURE 7.1: Bronchi sounds in time and time–frequency domains

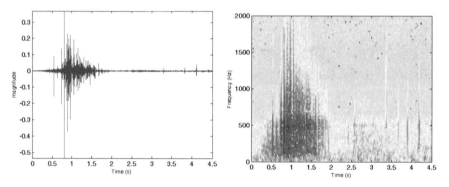

FIGURE 7.2: Fine crackles in time and time–frequency domains

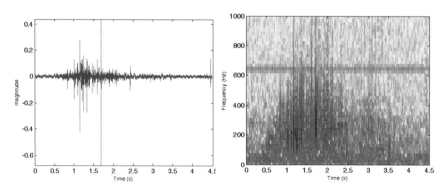

FIGURE 7.3: Coarse crackles in time and time–frequency domains

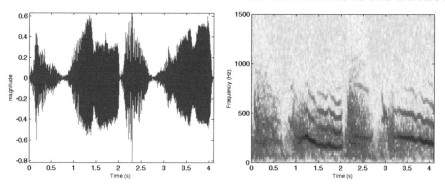

FIGURE 7.4: Wheeze sounds in time and time–frequency domains (two breaths)

Squawks.   Squawks that sometimes called squeaks are short wheezes usually heard in pneumonia. Fig. 7.5 shows a typical example of squawk sounds.

Rhonchi.   Rhonchi sounds are low-pitch sounds compared to wheezes and sound like snoring. They are often described as continuous sounds and usually clear after coughing. A typical rhonchi sound is shown in Fig. 7.6. These sounds are present in diseases in which the airway narrows.

Pleural rub sound.   These sounds are the sounds like running two pieces of leather against each other and are usually caused by pleural surface inflammation. A typical example is shown in Fig. 7.7.

Stridor sound.   These sounds are like wheezes but heard usually during inspiration while wheezes are heard during expiration. Stridor sounds are as a result of upper airway obstruction. A typical example is shown in Fig. 7.8.

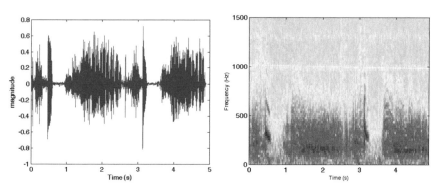

FIGURE 7.5: Squawk sounds in time and time–frequency domains (two breaths)

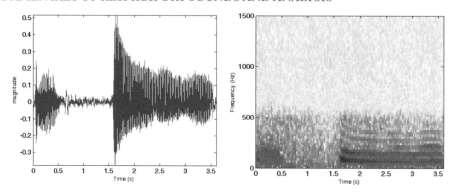

FIGURE 7.6: Rhonchi sounds in time and time–frequency domains (one breath)

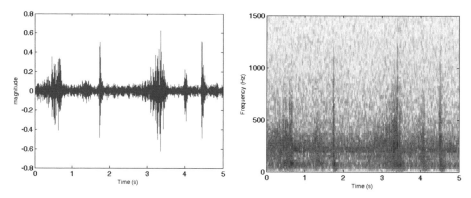

FIGURE 7.7: Pleural sounds in time and time–frequency domains (two breaths)

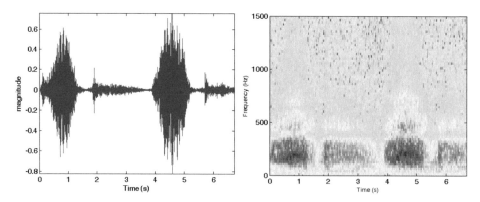

FIGURE 7.8: Stridor sounds in time and time–frequency domains (two breaths)

As one can note by the above description and figures, these symptomatic sounds have many similarities and usually a number of them are present together at the same respiratory condition. Therefore, one cannot expect a computer program to be developed that will detect with a high accuracy the respiratory condition only based on these sounds. Physicians look at many other symptoms along with these characteristic sounds to diagnose a respiratory condition. Even 100% accuracy in distinguishing these sounds from each other is not something to be expected from signal analysis of these sounds. However, in general, in order to aid the physicians to detect adventitious sounds and their locations and number of their occurrences over the period of recording, many research studies have tried to develop an automated detection algorithm for the symptomatic sounds and classify them as an aid for diagnosis. Most of the works have been focused on detecting wheezes and crackles. The main drawback in these studies is that the performance of the developed methods has been evaluated on data collected from only a few patients; hence, not all the cases of respiratory conditions were present in the data set. While detecting the adventitious lung sounds from the normal lung sounds is reasonably doable, distinguishing the adventitious sounds from each other, i.e., stridors from wheezes or squawks, with a high accuracy would be challenging. For example, none of the studies have tried to distinguish between the stridor and crackle sounds.

The techniques to detect adventitious sounds are more or less similar to those used for heart sound localization. It can be expected that wavelet analysis would result in better detection for crackle sounds as they are like spikes in the signal while wheezes can be detected simply by spectral analysis as they have a sinusoidal behavior. In essence, one should look to find the singular points of the time domain or frequency domain of the original lung sound signal by applying a threshold to the extracted feature of the signal. The techniques such as independent component analysis or blind source separation will not be effective in separating adventitious sounds from the each other as the source signals are not independent, for example, they are all originated by the turbulent flow in the airways.

In general, one may group the adventitious lung sounds into two classes of continuous and discontinuous sounds. Wheezes, rhonchi, bronchial, and stridor sounds belong to the first class, while crackles, squawks, and pleural rub sounds belong to the second class. The continuous sounds have a sinusoidal behavior and are seen as peaks in the successive spectra of the lung sound signal; hence, the most straightforward method is to apply a threshold to detect these peaks and combine it with some smart algorithm that uses the statistics of occurrence of these peaks to finally classify them as adventitious sounds versus vesicular sounds. To distinguish the sounds within each group requires another smart screening algorithm based on what is known about these sounds clinically as described briefly at the beginning of this section. However, almost all of the researches in this group have been focused on automatic detection of wheezes and not the rest of the sounds in the continuous class of adventitious sounds. One should note that even detection of wheezes from vesicular sounds when the wheezes are faint can become challenging and for sure automated classification of the sounds within the group would be very difficult. For a detailed description of the employed wheeze detection techniques, the interested reader may look at [67–71].

The second classes of adventitious sounds include the discontinuous sounds in the time domain signal and include fine and coarse crackles, pleural rubs and squawks. One may consider squawks to be in both groups as it has the characteristics of wheezes as well as crackles. One of the most effective methods in detection of discontinuous adventitious sounds is the use of multiresolution wavelet analysis to detect singularities, i.e., crackles. The interested reader may look at [72–75] for the details of the methods on crackle sounds detection.

CHAPTER 8

Acoustic Mapping and Imaging of Thoracic Sounds

Multichannel lung sound recording over the chest in the front and back has received much attention in the recent years due to its diagnostic potentials in a noninvasive manner. The conventional imaging systems for lungs are chest x-rays, CT, MRI that are well-established methods. The ultrasound method is fairly noninvasive; however, it has not been developed successfully due to the very high attenuation of high-frequency sounds at lung parenchyma.

Traditional auscultation by the stethoscope sounds much less reliable and valuable compared to the imaging methods mentioned above and also pulmonary function tests, pulmonary arteriography, radioisotope scanning, and biopsy of the lung that is used in some instances to make a definite diagnosis of the lung condition. However, auscultation remains as one of the most common devices used by physicians due to its simplicity, noninvasive manner, and availability. Indeed, in the situation of an acute respiratory distress in an emergency room, none of the above mentioned tests are likely to be used first. In such a situation, auscultation can be lifesaving.

Physicians, when listening to the lung sounds of a patient with a respiratory disease, often listen to the sounds over a few locations of the patient's chest both in the front and back. In fact by doing so, they gather information about the sounds coming from different parts of the lung and subjectively try to remember the quality of the sound at every location and assess them together for symptomatic sounds. Therefore, it makes their job much easier if there is a simultaneous multisensor recording over the chest wall so that the sounds can be listened to repeatedly. Furthermore, one can analyze the sounds in the time–frequency domain and have an acoustic image of the lung.

One of the first attempts on this topic was the development of a 16 electronic stethoscopes in a backpad (Fig. 8.1) that can be comfortably placed behind the patient and have the simultaneous digital sounds over the back of the chest wall stored in a computer for further analysis [67]. Over the years, the same group developed a software accompanying the multi-stethoscope jacket to display the sounds at every location and automatically count the number

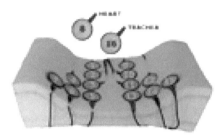

FIGURE 8.1: Backpad microphones. Adopted from [67] with permission

of symptomatic sounds such as crackles, strides, etc., as well as a three-dimensional (3D) image of the concentration of the symptomatic sounds over the chest so that the physician can detect the site affected by the disease objectively and also much faster.

Using simultaneous multisensor recordings of thoracic sounds from the chest wall, an acoustic imaging of the chest has recently been investigated for detecting plausible different patterns between healthy individuals and patients [76]. In that study, a new method for acoustic imaging of the respiratory system was developed and evaluated by a physical model of the lung as well as experimental data on four subjects and one patient. The sound speed, sound wavelengths at the frequencies of diagnostic values, and the geometry of the lung were taken into consideration in developing a model for acoustic imaging to provide a spatial representation of the intrathoracic sounds as opposed to the mapping of sounds on the thoracic surface. The acoustic imaging model was developed based on the calculation of a 3D data array [76],

$$s_k(t - |x_k - y|/c) = d^{|x_k - y|} \times r(y, t)/|x_k - y|^2,$$

where x_k ($k = 1, \dots, M$) are the positions of M microphones on the thoracic surface, $s_k(t)$ is the sound signal recorded at each microphone, y is the hypothetical source site, c is the sound speed, d is the uniform damping factor per unit length, and $r(y, t)$ is the signal emitted by the hypothetical source (intrathoracic sound). Having the signals recorded by M microphones, the resulted M equations can be solved by the least-squares fit to estimate $r(y, t)$, the hypothetical source. Each equation takes into account the delay between the hypothetical source and microphone (the left-hand side of the equation) and the geometric and assumed uniform damping on the source signal (the right-hand side of the equation).

The above model was tested on two subjects with 8 microphones and two other subjects with 16 microphones and one patient (a 5-year-old child with severe pneumonia) as well as on a physical model of the human lung. The results were congruent with the hypotheses that inspiratory sounds are produced predominantly in the periphery of the lung, while the expiratory sounds are produced more centrally [76]. However, it should be noted that the model was tested and evaluated only on four healthy subjects and the results with one of these four subjects was not consistent with the other three in terms of the concentration of the sound source on the upper right anterior portion of the chest wall, though this could be due to the fact that the number of microphones was not the same for all the four healthy subjects and the images obtained by 8 microphones showed substantial differences with those obtained by 16 microphones. In conclusion, based on the resulted images provided in [76], it seems that at least 16 microphones are needed for a reliable acoustic imaging of the chest.

Although, as the authors in [76] state, the acoustic imaging method is unlikely to compete with the CT, x-ray, or MRI techniques in terms of the information that they provide, the results are encouraging as it opens a new diagnostic possibility that upon further improvement in sensor technology, hardware design, and advances in computer technology to reduce the computational cost of the algorithm, may become a routine monitoring technique before using the other conventional methods, due to its noninvasive nature.

References

[1] J.L. Flanagan, *Speech Analysis Synthesis and Perception*, 2nd edition. New York: Springer, 1972.

[2] K. Ishizaka, J.C. French and J.L. Flanagan, "Direct determination of vocal tract wall impedance," *IEEE Trans. Acoust. Speech Signal Process.,* Vol. ASSP-23, pp. 370–373, 1975.doi:10.1109/TASSP.1975.1162701

[3] J.J. Fredberg and A. Hoenig "Mechanical response of the lungs at high frequencies," *J. Biomech. Eng.,* Vol. 100, pp. 57–66, 1978.

[4] G.R. Wodicka, K.N. Stevens, H.L. Golub, E.G. Cravalho and D.C. Shannon, "A model of acoustic transmission in the respiratory system," *IEEE Trans. Biomed. Eng.*, Vol. 36, No. 9, pp. 925–934, 1989.doi:10.1109/10.35301

[5] S.S. Kraman, "Speed of low-frequency sound through the lungs of normal men," *J. Appl. Physiol.*, Vol. 55, pp. 1862–1867, 1983.

[6] F.A. Duck, *Physical Properties of Tissue: A Comprehensive Reference Book*, 1st edition. San Diego, CA: Academic, 1990.

[7] P.J. Berger, E.M. Skuza, C.A. Ramsden and M.H. Wilkinson, "Velocity and attenuation of sound in the isolated fetal lung as it is expanded with air," *J. Appl. Physiol.*, Vol. 98, No. 6, pp. 2235–2241, 2005.doi:10.1152/japplphysiol.00683.2004

[8] A. Pohlmann, S. Sehati and D. Young, "Effect of changes in lung volume on acoustic transmission through the human respiratory system," *J. Physiol. Meas.*, Vol. 22, pp. 233–243, 2001.

[9] M. Mahagnah and N. Gavriely, "Gas density does not affect pulmonary acoustic transmission in normal men," *J. Appl. Physiol.*, Vol. 78, pp. 928–937, 1995.

[10] T. Bergstresser, D. Ofengeim, A. Vyshedskiy, J. Shane and R. Murphy, "Sound transmission in the lung as a function of lung volume," *J. Appl. Physiol.* Vol. 93, pp. 667–674, 2002.

[11] S. Lu, P.C. Doerschuk and G.R. Wodicka, "Parametric phase-delay estimation of sound transmitted through intact human lung," *Med. Biol. Eng. Comput.*, Vol. 33, No. 3, pp. 293–298, 1995. doi:10.1007/BF02510502

[12] H. Kiyokawa and H. Pasterkamp, "Volume-dependent variations of regional lung sound, amplitude and phase," *J. Appl. Physiol.*, Vol. 93, pp. 1030–1038, 2002.

[13] H. Pasterkamp, S.S. Kraman and G.R. Wodicka, "Respiratory sounds. Advances beyond the stethoscope," *Am. J. Respir. Crit Care Med.*, Vol. 156, No. 3, Pt 1, pp. 974–987, Sept. 1997.

[14] Z. Moussavi, M.T. Leopando, H. Pasterkamp and G. Rempel, "Computerized Acoustical Respiratory Phase Detection without Airflow Measurement," *Med. Biol. Eng. Comput.*, Vol. 38, No. 2, pp. 198–203, 2000.doi:10.1007/BF02344776

[15] S.C Tarrant, R.E Ellis, F.C Flack and W.G. Selley, "Comparative review of techniques for recording respiratory events at rest and during deglutition," *J. Dysphagia*, Vol. 12, pp. 24–38, 1997.

[16] G. Soufflet, G. Charbonneau, M. Polit, P. Attal, A. Denjean, P. Escourrou and C. Gaultier, "Interaction between tracheal sound and flow rare: a comparison of some different flow evaluations from lung sounds," *IEEE Trans. Biomed. Eng.*, Vol. 37, No. 4, pp. 384–391, 1990.

[17] I. Hossain and Z. Moussavi, "Respiratory airflow estimation by acoustical means," in *Proc. IEEE 2nd joint EMBS/BMES Conf.*, Houston, TX, 2002, pp. 1476–1477.

[18] Y. Yap and Z. Moussavi, "Acoustic airflow estimation from tracheal sound power," in *Proc. IEEE Canadian Conf. Elec. Comp. Eng. (CCECE)*, 2002, pp. 1073–1076.

[19] C. Que, C. Kolmaga, L. Durand, S. Kelly and P. Macklem, "Phonospirometry for non-invasive measurement of ventilation: methodology and preliminary results," *J Appl. Phys.*, Vol. 93, pp. 1515–1526, 2002.doi:10.1063/1.1428106

[20] A. Yadollahi and Z. Moussavi, "A robust method for estimating respiratory flow using tracheal sound entropy," *J. IEEE Trans. Biomed. Eng.*, Vol. 53, No. 4, pp. 662–668, 2006.

[21] N. Gavriely and D. Cugell, "Airflow effects on amplitude and spectral content of normal breath sounds," *Am. Physiol. Soc.*, pp. 5–13, 1996.

[22] M. Golabbakhsh, "Tracheal breath sound relationship with respiratory flow: Modeling, the effect of age, and airflow estimation," M.Sc. thesis, Electrical and Computer Engineering Department, University of Manitoba, 2004.

[23] A. Yadollahi and Z. Moussavi, "Robust method for heart sound localization using lung sounds entropy," *J. IEEE Trans. Biomed. Eng.*, Vol. 53, No. 3, pp. 497–502, 2006.

[24] L.J. Hadjileontiadis and S.M. Panas, "A wavelet-based reduction of heart sound noise from lung sounds," *Int. J. Med. Inform.*, Vol. 52, pp. 183–190, 1998.

[25] I. Hossain and Z. Moussavi, "An overview of heart-noise reduction of lung sound using wavelet transform based filter," in *Proc. 25th IEEE Eng. Med. Biol. Soc. (EMBS)*, 2003, pp. 458–461.

[26] V.K. Iyer, P.A. Ramamoorthy, H. Fan and Y. Ploysongsang, "Reduction of heart sounds from lung sounds by adaptive filtering," *J. IEEE Trans. Biomed. Eng.*, Vol. 33, No. 12, pp. 1141–1148, 1986.

[27] L. Yip and Y.T. Zhang, "Reduction of heart sounds from lung sound recordings by automated gain control and adaptive filtering techniques," in *Proc. 23rd IEEE Eng. Med. Biol. Soc. (EMBS)*, 2001, pp. 2154–2156.

[28] L. Guangbin, C. Shaoqin, Z. Jingming, C. Jinzhi and W. Shengju, "The development of a portable breath sounds analysis system," in *Proc. 14th Ann. Int. Conf. IEEE EMBS*, 1992, pp. 2582–2583.

[29] L. Yang-Sheng, L. Wen-Hui and Q. Guang-Xia, "Removal of the heart sound noise from the breath sound," in *Proc. 10th IEEE Eng. Med. Biol. Soc. (EMBS)*, 1988, pp. 175–176.

[30] M. Kompis and E. Russi, "Adaptive heart-noise reduction of lung sounds recorded by a single microphone," in *Proc. 14th Ann. Int. Conf. IEEE EMBS*, 1992, pp. 691–92.

[31] Z. Moussavi, et al., "Screening and adaptive segmentation of vibroarthrographic signals," *J. IEEE Trans. Biomed. Eng.*, Vol. 43, No. 1, 15–23, 1996.

[32] J. Gnitecki, I. Hossain, Z. Moussavi and H. Pasterkamp, "Qualitative and quantitative evaluation of heart sound reduction from lung sound recordings," *J. IEEE Trans. Biomed. Eng.*, Vol. 52, No. 10, pp. 1788–1792, Oct. 2005.

[33] T.W. Lee, *Independent Component Analysis: Theory and Applications,* 2nd edition. Dordrecht: Kluwer, 2000, chapters 2 and 4.

[34] N. Murata, S. Ikeda and A. Ziehe, "An approach to blind source separation based on temporal structure of speech signals," *Neurocomputing*, Vol. 41, pp. 1–24, 2001.

[35] M. Pourazad, Z. Moussavi, F. Farahmand and R. Ward, "Heart sounds separation from lung sounds using independent component analysis," in *Proc. IEEE Eng. Med. Biol. Soc. (EMBS)*, Sept. 2005.

[36] M.T. Pourazad, Z.K. Moussavi and G. Thomas, "Heart sound cancellation from lung sound recordings using adaptive threshold and 2D interpolation in time–frequency domain," in *Proc. IEEE Eng. Med. Biol. Soc. (EMBS)*, Sept. 2003, pp. 2586–2589.

[37] M.T. Pourazad, Z. Moussavi and G. Thomas, "Heart sound cancellation from lung sound recordings using time–frequency filtering," *J. Med. Biol. Eng.*, Vol. 44, No. 3, pp. 216–225, 2006.

[38] Z. Moussavi, D. Floras and G. Thomas, "Heart sound cancellation based on multiscale products and linear prediction," in *Proc. 26th IEEE Eng. Med. Biol. Soc. (EMBS)*, Sept. 2004, pp. 3840–3843.

[39] D. Floras, Z. Moussavi and G. Thomas, "Heart sound cancellation based on multiscale product and linear prediction," *J. IEEE, Trans. Biomed. Eng.*, 2006, to be published.

[40] J. D. Proakis and D. G. Manolakis, *Digital Signal Processing.* Englewood Cliff, NJ: Prentice-Hall, 1996.

[41] W. Etter, "Restoration of a discrete-time signal segment by interpolation based on the left-sided and right sided autoregressive parameters," *J. IEEE Trans. Signal Process.*, Vol. 44, No. 5, pp. 1124–1135, May 1996.doi:10.1109/78.502326

[42]   C. Ahlstrom, O. Liljefeldt, P. Hult and P. Ask, "Heart sound cancellation from lung sound recordings using recurrence time statistics and nonlinear prediction," *J. IEEE Signal Process. Lett.*, Vol. 12, No. 12, pp. 812–815, 2005.

[43]   L. Hadjileontiadis and S. Panas, "Adaptive reduction of heart sounds from lung sounds using forth-order statistics," *J. IEEE Trans. Biomed. Eng.*, Vol. 44, No. 7, pp. 642–648, 1997.

[44]   J. Gnitecki and Z. Moussavi, "Variance fractal dimension trajectory as a tool for heart sound localization in lung sounds recordings," in *Proc. IEEE Eng. Med. Biol. Soc. (EMBS)*, pp. 2420–2423, 2003.

[45]   J. Gnitecki, Z. Moussavi and H. Pasterkamp, "Recursive least square adaptive noise cancellation filtering for heart sound in lung sounds recording," in *Proc. IEEE Eng. Med. Biol. Soc. (EMBS)*, pp. 2416–2419, 2003.

[46]   M.T. Pourazad, "Heart sounds reduction from lung sounds recordings applying signal and image processing techniques in time–frequency domain," M.Sc. thesis, Electrical and Computer Engineering Department, University of Manitoba, 2004.

[47]   W. Kinsner, "Batch and real-time computation of a fractal dimension based on variance of a time series," Technical Report, DEL94-6, Dept. of Electrical & Computer Eng., University of Manitoba, Winnipeg, Manitoba, Canada, June 1994.

[48]   P. Bao and L. Zhang, "Noise reduction for magnetic resonance images via adaptive multiscale products thresholding" *J. IEEE Trans. Med. Imaging,* Vol. 22, No. 9, pp. 1089–1099, 2003.

[49]   D. Flores, G. Thomas and M.C. Phelan, "Clutter Reduction of GPR Images Using Multiscale Products," in *Proc. Int. Conf. Antennas, Radar and Wave Propagation IASTED 2004*, Banff, Canada, July 2004.

[50]   A. Papoulis, *Probability, Random Variables and Stochastic Processes*. New York: McGraw-Hill, 1991.

[51]   B. Silverman, *Density Estimation for Statistics and Data Analysis*. New York: Chapman and Hall, 1986.

[52]   H. Pasterkamp, "Acoustic markers of airway responses during inhalation challenge in children," *Pediatr. Pulmonol. Suppl.*, Vol. 26, pp. 175–176, 2004.

[53]   H. Pasterkamp, R. Consunji-Araneta, Y. Oh and J. Holbrow, "Chest surface mapping of lung sounds during methacholine challenge," *Pediatr. Pulmonol.*, Vol. 23, No. 1, pp. 21–30, 1997.doi:10.1002/(SICI)1099-0496(199701)23:1<21::AID-PPUL3>3.0.CO;2-S

[54]   J. Gnitecki, Z. Moussavi and H. Pasterkamp, "Classification of lung sounds during bronchial provocation using waveform fractal dimensions," in *Proc. 26th IEEE Eng. Med. Biol. Soc. (EMBS)*, Sept. 2004, pp. 3844–3847.

[55] E. Conte, A. Vena, A. Federici, R. Giuliani and J.P. Zbilut, "A brief note on possible detection of physiological singularities in respiratory dynamics by recurrence quantification analysis of lung sounds," *J. Chaos Soliton Fract.*, Vol. 21, pp. 869–877, 2004.

[56] M. Aboofazeli and Z. Moussavi, "Analysis of normal swallowing sounds using nonlinear dynamic metric tools," in *Proc. 26th IEEE Eng. Med. Biol. Soc. (EMBS)*, 2004, pp. 3812–3815.

[57] M. Banbrook, S. McLaughlin and I. Mann, "Speech characterization and synthesis by nonlinear methods," *J. IEEE Trans. Speech Audio Process.*, Vol. 7, No.1, pp. 1–17, 1999.

[58] J. Gnitecki and Z. Moussavi, "The fractality of lung sounds: a comparison of three waveform fractal dimension algorithms," *J. Chaos Solitons Fractals*, Vol. 26(4), pp. 1065–1072, 2005.

[59] M.J. Katz, "Fractals and the analysis of waveforms", *J. Comput. Biol. Med.*, Vol. 18, No. 3, pp. 145–156, 1988.

[60] H.D.I. Abarbanel, R. Brown, J.J. Sidorowich and L. Tsimring, "The analysis of observed chaotic data in physical systems," *Rev. Mod. Phys.*, Vol. 65, No. 4, pp. 1331–1392, 1993.doi:10.1103/RevModPhys.65.1331

[61] M.T. Rosenstein, J.J. Collins and C.J. De Luca," Reconstruction expansion as a geometry-based framework for choosing proper delay times," *Physica D*, Vol. 73, pp. 82–98, 1994.doi:10.1016/0167-2789(94)90226-7

[62] J.P. Zbilut, J.M. Zaldivar-Comenges and F. Strozzi, "Recurrence quantification based on Lyapunov exponents for monitoring divergence in experimental data," *Phys. Lett. A*, Vol. 297, pp. 173–181, 2002.

[63] R. C. Hilborn, "*Chaos and Nonlinear Dynamics*", 2nd edition. Oxford : Oxford University Press, 2000.

[64] A. Wolf, J.B. Swift, H.L. Swinney and J.A. Vastano, "Determining Lyapunov exponents from a time series," *Physica D*, Vol. 16, pp. 285–317, 1985.

[65] J. Gnitecki, Z. Moussavi, and H. Paskterkamp, "Geometrical and dynamical state space parameters of lung sounds," in *5th Int. Workshop on Biosignal Interpretation (BSI)*, Sept. 2005, pp. 113–116.

[66] C. Ahlstrom, A. Johansson, P. Hult and P. Ask, "Chaotic dynamics of respiratory sounds," *J. Chaos Solitons Fractals*, Vol. 29, pp. 1054–1069, 2006.

[67] R.L.H. Murphy, "Localization of chest sounds with 3D display and lung sound mapping." U.S. Patent 5,844,997, Dec. 1, 1998.

[68] S.A. Tapildou, L.J. Hadjileontiadis, I.K. Kitsas, K.I. Panoulas, T. Penzel, V. Gross and S.M. Panas, "On applying continuous wavelet transform in wheeze analysis," in *Proc. 26th IEEE Eng. Med. Biol. Soc. (EMBS)*, Vol. 2, pp. 3832–3835, 2004.

[69] Y. Shabtai-Musih, J.B. Grotberg and N. Gavriely, "Spectral content of forced expiratory wheezes during air, He, and SF6 breathing in normal humans," *J. Appl. Physiol.*, Vol. 72, pp. 629–635, 1992.

[70] T.R. Fenton, H. Pasterkamp, A. Tal and V. Chernick, "Automated spectral characterization of wheezing in asthmatic children," *J. IEEE Trans. Biomed. Eng.*, Vol. 32, pp. 50–55, 1985.

[71] R.J. Riella, P. Nohama, R.F. Borges and A.L. Stelle, "Automatic wheezing recognition in recorded lung sounds," in *Proc. IEEE Eng. Med. Biol. Soc. (EMBS)*, 2003, pp. 17–21.

[72] J. Hoevers and R.G. Loudon, "Measuring crackles," *J. Chest*, Vol. 98, pp. 1240–1243, 1990.

[73] S.K. Holford, "Discontinuous adventitious lung sounds: measurement, classification and modeling," Sc.D. thesis, Massachusetts Institute of Technology, Cambridge, MA, 1982.

[74] Y. Maeda, K. Nitta, Y. Yui, T. Shida, K. Wagatsuma and T. Muraoka, "A classification of crackles by the linear predictive coding method," *Nippon. Kyobu Shikkan Gakkai Zasshi*, Vol. 26, pp. 587–593, 1988.

[75] L. Hadjileontiadis and S. Panas, "Separation of discontinuous adventitious sounds from vesicular sounds using a wavelet-based filter," *J. IEEE Trans. Biomed. Eng.*, Vol. 44, No. 12, pp. 1269–1281, 1997.

[76] M. Kompis, H. Paskterkamp and G. Wodicka, "Acoustic imaging of the human chest," *J. Chest*, Vol. 120, No. 4, pp. 1309–1321, 2001.

Biography

Zahra M. K. Moussavi received her B.Sc. from Sharif University of Technology, Iran in 1987, M.Sc. from the University of Calgary in 1993, and Ph.D. from University of Manitoba, Canada in 1997, all in Electrical Engineering. She then joined the respiratory research group of the Winnipeg Children's Hospital and worked as a research associate for 1.5 years. In 1999, she joined the Biomedical Engineering Department of Johns Hopkins University as a postdoctoral fellow for one year. Following that, she joined the University of Manitoba, Department of Electrical and Computer Engineering as a faculty member, where she is currently an associate professor. She is also an adjunct professor at the TR lab of Winnipeg. Her current research includes respiratory and swallowing sound analysis, postural control and balance, rehabilitation and human motor learning. Dr. Moussavi is a senior member of IEEE, Engineering in Medicine and Biology (EMBS), member of CMBES and International Lung Sound (ILSA) associations. She is also currently the EMBS Chapter Chair, Winnipeg Section.

Printed in the United States
by Baker & Taylor Publisher Services